Good Governance and Development

Challenges in India

Good Governance and Development

Challenges in India

Edited by

C. Vinodan

Co-ordinator, Rajiv Gandhi Chair for Public Policy Studies
and
Honorary Director, Institute for Contemporary Chinese Studies (ICCS), School of International Relations and Politics, Mahatma Gandhi University, Kottayam, Kerala

New Century Publications
New Delhi, India

NEW CENTURY PUBLICATIONS
4800 / 24, Bharat Ram Road,
Ansari Road, Daryaganj,
New Delhi - 110 002 (India)

Tel.: 011-2324 7798, 4358 7398, 4101 7798
E-mail: indiatax@vsnl.com • info@newcenturypublications.com
www.newcenturypublications.com

Editorial office:
LG – 7, Aakarshan Bhawan,
4754-57/23, Ansari Road, Daryaganj,
New Delhi – 110 002

Tel.: 011-4356 0919

First Published: **2017**

ISBN: **978-81-7708-446-7**

Published by New Century Publications and printed at Milan Enterprises, New Delhi.

Designs: Patch Creative Unit, New Delhi.

PRINTED IN INDIA

About the Book

Governance is a theoretical concept referring to the actions and processes by which stable practices and organizations arise and persist. These actions and processes may operate in formal and informal organizations of any size; and they may function for any purpose. Conceiving of governance in this way, one can apply the concept to states, corporations, NGOs, partnerships, and any number of humans engaged in some purposeful activity.

The World Bank defines governance as "the manner in which power is exercised in the management of a country's economic and social resources for development".

Better governance is crucial for translating the large development and welfare expenditure of the public authorities into enduring outcomes on the ground. While there is general appreciation that development programmes have the right objectives, their implementation on the ground is poor. Implementation of programmes can be improved through a multi-faceted approach relying on professionalization of public service delivery, total quality management (TQM), and innovative use of digital and other technologies which improve monitoring and supervision. It can also be improved through: (a) greater emphasis on social mobilisation and capacity building, (b) strengthening of local institutions and (c) building deeper partnerships with civil society organisations and the community to determine the needs and aspirations of the people.

The present volume contains 11 articles—authored by experts in the field of public policy—that provide deep insights into various aspects of governance in India in the context of socio-economic development.

About the Editor

Dr. C. Vinodan is currently Co-ordinator, Rajiv Gandhi Chair for Public Policy Studies and Honorary Director, Institute for Contemporary Chinese Studies (ICCS), School of International Relations and Politics, Mahatma Gandhi University, Kottayam, Kerala. He has worked previously at the Post-graduate and Research Department of Political Science, Sree Narayana College, Kollam and Department of Political Science, University of Kerala.

He obtained his M.A. (II Rank), M.Phil. and Ph.D. degrees in Political Science from the University of Kerala. He was awarded the Visiting Scholar Fellowship (2003) by the Claremont Graduate University, California, USA. He has attended numerous courses in international relations in India and abroad. These include the 2016 Sino-Indian Cross-Cultural Dialogues held by the Institute of Advanced International Studies, Sun Yat-sen University, Guangzhou, China; and the US Department of States Fulbright American Studies Institute (FASI) course on foreign policy at the University of Delaware, Philadelphia in 2000.

Dr. Vinodan has also attended and presented papers at various national and international conferences/seminars. He is also the recipient of a number of awards including Kerala University Post-Doctoral Award (2003-2005). He is member of the editorial board of the *Journal of Human Security*, *Indian Journal of Politics and International Relations* and *South Asian Journal of Diplomacy*. He has published 40 research papers and 4 books.

His areas of teaching and research activities include international relations, security studies, Chinese studies, South Asian studies, energy, environment and human security.

Contents

Contributors

Tej Pratap Singh	Professor, Department of Political Science, Banaras Hindu University, Varanasi.
Chandrakant Puri	Rajiv Gandhi Chair Professor, University of Mumbai, Mumbai.
Kuldip Singh	Professor, Department of Political Science, Punjabi University, Punjab.
Sukhjit Singh	Research Scholar, Department of Political Science, Punjabi University, Punjab.
K.C. Sreekumar	Associate Professor of Political Science, NSS Law College, Kollam.
Haans J. Freddy	Assistant Professor, Department of Political Science, Madras Christian College, Chennai.
Rajeev M.M.	Assistant Professor, Department of Social Work, Central University of Rajasthan, Ajmer.
Anju Liz Kurian	UGC Post-doctoral Fellow, School of International Relations and Politics, Mahatma Gandhi University, Kottyam.
C. Vinodan	Co-ordinator, Rajiv Gandhi Chair for Public Policy Studies, School of International Relations and Politics, Mahatma Gandhi University, Kottyam.
K.M. Seethi	Professor and Chairman, Centre for Cross-National Communication in South Asia ,Co-ordinator, KPS Menon Chair for Diplomatic Studies, School of International Relations and Politics, Mahatma Gandhi University, Kottayam, Kerala

Praveen Kumar Associate Professor, Centre for Political Studies, Central University of South Bihar, Gaya.

Sijo K. Manuel Assistant Professor, Department of Political Science, St. Thomas College, Kottayam.

Preface

In its most abstract sense, governance is a theoretical concept referring to the actions and processes by which stable practices and organizations arise and persist. These actions and processes may operate in formal and informal organizations of any size; and they may function for any purpose, good or evil, for profit or not. Conceiving of governance in this way, one can apply the concept to states, corporations, NGOs, partnerships, and any number of humans engaged in some purposeful activity.

The World Bank defines governance as "the manner in which power is exercised in the management of a country's economic and social resources for development".

A variety of entities (known generically as *governing bodies*) can govern. The most formal is a *government*, a body whose sole responsibility and authority is to make binding decisions in a given geo-political system (such as a state) by establishing laws. Other types of governing bodies are possible. These include an organization (such as a corporation recognized as a legal entity by a government), a socio-political group (chiefdom, tribe, family, religious denomination, etc.), or another, informal group of people.

Whatever form the entity takes, its *governance* is the way the rules, norms and actions are produced, sustained, regulated and held accountable. The degree of formality depends on the internal rules of a given organization. As such, governance may take many forms, driven by many different motivations and with many different results. For instance, a government may operate as a democracy where citizens vote on who should govern and the public good is the goal, while a non-profit organization may be governed by a small board of directors and pursue more specific aims.

In addition, a variety of external actors without decision-making power can influence the process of governing. These include lobbies, political parties, and the media.

Better governance is crucial for translating the large

development and welfare expenditure of the public authorities into enduring outcomes on the ground. While there is general appreciation that development programmes have the right objectives, their implementation on the ground is poor. Implementation of programmes can be improved through a multi-faceted approach relying on professionalization of public service delivery, total quality management (TQM), innovative use of digital and other technologies which improve monitoring and supervision. It can also be improved through: (a) greater emphasis on social mobilisation and capacity building, (b) strengthening of local institutions and (c) building deeper partnerships with civil society organisations and the community to determine the needs and aspirations of the people.

Implementation in many areas—particularly in infrastructure development involving large projects—is held up for a variety of reasons. Co-ordination needed across different agencies to facilitate progress in project implementation is often lacking and can lead to long delays and cost over-runs. Project management capabilities must be improved for the country to get better returns from public investment in infrastructure and also in the social sectors.

A number of initiatives need to be urgently pursued to rid the system of corruption, which is both morally abhorrent and imposes economic costs. Several legislative measures are needed. These include the establishment of an effective *Lokpal,* introduction of a law on public procurement and transparency, and the creation of a legislative framework governing the functioning of regulatory institutions so as to ensure both functional autonomy and accountability.

To combat corruption, it is imperative to ensure speedy prosecution and trial in corruption cases. The long delays in the judicial process are an important factor behind the growing cynicism about the rule of law in our system. Reforms in the legal process need to be put in place without further delays.

Corruption distorts the decision-making mechanism and leads to an inefficient distribution of resources. Improving anti-corruption efforts is one of the highest-rated priorities of the business leaders. India should work closely with relevant international organizations and give particular importance to the fight against corruption in the public and private sector. Measures to increase transparency in government-business relations will have a corrective impact on corruption.

Information technology (IT) is the world's fastest growing economic activity. It is transforming resource-based economies to knowledge-based economies. IT has become the greatest agent of change and promises to play this role even more dramatically in future. IT has changed every aspect of human life, apart from impacting changes in the field of communications, trade, manufacturing, services, culture, entertainment, education, research and national security. IT has broken old barriers and is building new interconnections in the emerging concept of a single global village. It has also become one of the critical indicators of the progress of nations, communities and individuals.

The advent of IT offers opportunities to overcome historical disabilities. IT is a tool that will enable nations to achieve the goal of becoming a strong, prosperous and self-confident state. IT promises to compress the time it would otherwise take for countries to advance rapidly in their march towards faster development.

Presently, various services are provided by the Central and State Governments. These include ration cards, caste certificates, income certificates, certificates for proof of residence, passports and similar other services. It is important that these are available within a prescribed time line. Failures and deficiencies in delivery of public services lead to dissatisfaction and public anger against the government. Various government agencies have taken initiatives to notify services and time-limits within which these are to be availed of. An important aid to delivery of services can be the use of e-governance and technology. The use of technology in delivery public services needs to be expanded rapidly to reduce delays and discretion.

The present volume contains 11 research articles, authored by experts in the field of public policy, which provide deep insights into various aspects of governance in India. I place on record my deep sense of gratitude to all the scholars for their worthy contributions to this volume. Their views on governance in the context or development would surely add, constructively, to the existing body of literature on the subject.

2017 **C. Vinodan**

1

Maoist Insurgency

Tej Pratap Singh

Former Prime Minister Manmohan Singh described Maoist insurgency as the single largest threat to internal security in 2004. Since then more than a decade has passed but threat has not diminished. Unlike ethnic conflict in the North-East and religious extremism in Kashmir valley, Maoist insurgency is not separatist, constraining Government in the use of armed forces to crush Maoists as they are underprivileged and indoctrinated poor fellow citizens of the nation. Maoists are hell bent upon overthrowing the government and capturing political power at Delhi through protracted armed struggle forgetting the fact that they are confronting the fourth most powerful armed forces of the world after US, China and Russia. Government since the eruption of insurgency has been by and large treating it as a law and order issue and trying to crush it through the use of disproportionate force. Since the root causes of Maoism have not been addressed, the force has given only temporary respite to the government.

Left wing extremism (LWE) in India popularly known as Maoism and Naxalism has been caused mainly because of governance deficit, democratic deficit and development deficit or rather because of the triple crises of governance, democracy and development. All three deficits and crises are interlinked and form a vicious triangle, which has led to the formation of violence triangle, where the triple deficits/crises trigger apathy among the vast sections of society in the midst of hopelessness and helplessness and also violent/aggressive behaviour by the tiny few who take up arms and challenge the whole system created by the state with the tacit and active support of the masses. Since 'state is a human community that (successfully) claims the monopoly of the legitimate use of physical force within a given territory' (Max Weber, 1919) [1] and so no state can allow the use of force by the non-state actors (Maoists) in the name of redistribution of wealth or for the abolition of inequality on the basis of any ideology or dogma.

The result is the spiral of violence between the state and Maoists, which has spatial and human implications. Both are engaged in a fierce competition to bring more and more geographical terrain and people under their influence and control. Both are engaged into area domination strategy to wrest the control of more and more geographical territory and win the hearts and minds of people to succeed in the ultimate objective of decimating each other completely. This is basically a political war with political objectives but fought with military means.

Since it is insurgency and so spatial factors though significant, become secondary to human factor. Rather than wresting the territory from the state, insurgents aim at winning the people to their side. Whoever succeeds in winning the people to their side will also succeed in gaining control of the territory too. It is not a conventional inter-state war but intra-state unconventional war being waged between state and non-state actors (insurgents). It is a low intensity guerrilla war, where combatants are state's own citizens. People are challenging/resisting their own state, which has created a civil war type of situation. Consequently, the state finds its hands tied as they are fighting their own people (citizen). State cannot resort to disproportionate use of force as it will alienate its own people and undermine its legitimacy. It will lead to enormous collateral damage and fuel insurgency further.

Till this date, Indian state has not used its armed forces in its fight against Maoists. Indian armed forces have been used extensively in Kashmir and North-East to suppress insurgency. The rationale behind the use of armed forces in these two regions of Indian Union is that they are separatists, whereas Maoists are not. However, the new NDA government led by

Narendra Modi is revising the policy of the UPA government to tackle Maoists (The Telegraph, October 2014). [2]

Neo-liberalism and Maoist Insurgency

There is a direct correlation between adoption of neo-liberal policies by India in the wake of globalisation, liberalisation and privatisation and rise of Maoist insurgency in India. Indira Gandhi initiated the liberalisation process after her return to power in 1980 and Peoples War Group (PWG) formed in 1980 undertook the task of reviving the Maoist movement from Telangana region of Andhra Pradesh. As the liberalisation process accelerated during Rajiv Gandhi's regime, Maoists also began consolidating in Andhra Pradesh and Bihar. During 1990s with full-fledged embrace of Washington consensus and neo-liberal policies under the duo of Narsimha Rao-Manmohan Singh, Maoists also rapidly increased their area of influence (Murzban Zal, 2014). [3]

Globalisation, liberalisation and privatisation led to rapid but uneven economic growth where a tiny section of the population benefited whereas a large section of the population was excluded from the growth process. The exclusion of a vast number of people from the growth process created a lot of resentment and discontent amongst the masses, swelling the ranks of Maoists. India was divided into 'shining India' [4] and 'suffering India'. [5] Inequality increased and gulf between the rich and poor widened. During this phase of neo-liberalism, Maoists recorded rapid expansion and spread to one third of the country by the time the architect of neo-liberalism in India, Manmohan Singh transited from Finance Ministry to the office of Prime Minister in 2004. While addressing Chief Ministers of Naxal affected states, PM Manmohan Singh on April 13, 2006 observed, 'it would not be an exaggeration to say that the problem of Naxalism is the single biggest internal security challenge ever faced by our country'. [6]

With the advent of neo-liberalism, the public sector ceased to enjoy the 'commanding heights of the economy' [7] and the role of private sector increased. Disinvestment led to the privatisation of public sector enterprises or dilution of state control. Natural resources owned by the state were auctioned or allocated to the corporate houses arbitrarily, resulting in enormous loss to the state exchequer. All this led to ushering in the era of crony capitalism and massive corruption. Instead of protecting the interests of poor people, particularly *adivasis*, and poor peasants, the State was often perceived to be protecting the interest of corporate houses thus eroding the faith of poor adivasis and other marginalised people.

Corporate interests expanded and deepened in mining and other spheres of the economy. Minerals lie buried in the areas inhabited by the adivasis. Corporate houses needed large chunks of land for mining and to establish power plants, factories, etc. The State facilitated the acquisition of land by corporates for mining and industrial purposes in tribal areas which resulted in displacement of the adivasis. This was described by a Committee set up the Ministry of Rural Development as 'the biggest grab of tribal lands after Columbus' (Committee on State Agrarian Relations and Unfinished Task of Land Reforms, Draft Report, New Delhi, 2009, 160). [8] This pitted corporate houses against the adivasis. Extraction of minerals from tribal areas polluted their water sources, made their land infertile and destroyed their habitats and sources of livelihood resulting into alienation of adivasis from the state. This provided a fertile ground to Maoists to exploit the situation to their own advantage. They organised the adivasis against the state and transformed the tribal areas of India as 'the heartland of rebellion' (Navlakha, 2012). [9]

Theoretical Perspectives

Various theoretical frameworks have been used for understanding and explaining LWE in India. However, most popular among them is the greed and grievance theory (Collier and Hoeffler). [10] This theory has been developed in binary terms of either-or (either greed or grievance). Greed theory suits the state and grievance theory the insurgents (Maoists).

Before going beyond these two theories, it would be pertinent to explain these two theories and their application for the understanding of the Maoist violence. Greed theory subscribed by the state postulates that insurgents (Maoists) are a bunch of criminals, who are using ideology as a mask to conceal their true criminal face. They use violence to intimidate the people and create an atmosphere of fear. Their ultimate objective is to make money and lead a comfortable life. They develop vested interests in insurgency and want its continuance. Insurgents do not want an end to insurgency as it has become a source of livelihood for them. It is fashionable in insurgency ravaged area for the youths to pick up guns and become rebels. Ideology is a smokescreen, which provides legitimacy to their violence in the name of redistribution of wealth and social justice.

State treats insurgents as hardened criminals and extortionists and LWE as law and order problem and advocates the use of violence or counter-violence (counter-insurgency) to put an end to insurgency. Indian state is no exception and so from the very beginning of Naxalism in the late 1960s, the state has been resorting to disproportionate use of force to crush naxalism. Indian state succeeded initially to crush naxalism in 1970s. With the death of the top Naxal leader and their chief ideologue Charu Majumdar in 1972 in police custody and killing or arrest of top leaders like Kanu Sanyal, Jangal Santhal, etc., it appeared that naxal movement has fizzled out. But since the root causes of Naxalism were not addressed, therefore it bounced back with vengeance in 1980s and since then it has been spreading without any let up. Now if we go by the data provided by the Ministry of Home Affairs through RTI, one third of the country, i.e. around 200 districts in India are affected by Maoism. [11]

Out of these, more than 30 districts are worst affected and for these LWE affected districts, the Government of India (GOI) has launched an Integrated Action Plan (IAP) programme to address the issue of development deficits. Since these regions suffer from security deficits caused by development deficit, hence IAP is supplemented by Security Related Expenditure (SRE) [12] scheme too. IAP and SRE have been specially designed to address the issue of LWE.

The rival theory of greed is grievance theory of violence. Insurgents have genuine socio-economic grievances against the society and the polity. Since their grievances have not been addressed satisfactorily by the state; so these people have been left with no option but to pick up arms and rebel against the state. Naxalism/Maoism has provided ideological justification to their armed struggle against the state. Maoists argue that protracted armed struggle is the only way to redress the grievances of impoverished, oppressed and exploited people. Maoism is spreading and deepening only in those regions of the country, which are most backward and poverty stricken. The so-called red-corridor in the words of media and compact revolutionary zone (CRZ) in the words of Maoists, where Maoists are running or striving to run parallel government and intend to oust state completely is characterised by the triple deficits of governance, development and democracy.

In more than seven decades of Independence, there has been no significant change in the lives of people living in this vast stretch of land extending from Nepal in the north to Andhra Pradesh in the deep South. Former home minister L.K. Advani termed it as ‘from Pashupati (Nepal) to Tirupati (Andhra) [13] or Kathmandu to Kudremukh (Karnataka)’. Of late, Maoist violence has been reported in Kerala too. All these areas are forested, hilly and inaccessible, mainly inhabited by the indigenous people, who live a wretched life. Maoists have taken up the grievances of the oppressed people who in the Indian contexts are mainly adivasis (indigenous people) and dalits (lower caste people). These two social groups constitute the main support base of Maoists. The day state will succeed in winning these two social groups viz. adivasis and dalits to its side by addressing their grievances, the Maoist insurgency will come to an end.

Beyond Greed and Grievance

Theoretical framework of greed and grievance for explaining Maoist insurgency is an oversimplification of complex issue of naxalism/Maoism in India. Both the theories have some elements of truth but both fail to offer satisfactory explanation of Maoist insurgency. Criminalisation has crept into the movement and many criminals have joined the movement, who have nothing to do with ideology justifying the greed theory. This is particularly true in the case of Jharkhand (Singh, 2013) [14], where dozens of Maoist outfits with different names are primarily extortionists and indulge in all kinds of criminal activities. Most of the splinter groups are basically vigilant groups and indulge in serious human rights abuse. These splinter Maoist groups are engaged in fierce turf war with other splinter groups and parent party.

But to paint all the insurgents with the same brush will be unfair to the ideologically committed insurgents, who have sacrificed everything for the sake of their belief and ideology. They can be anything but not greedy or criminal. Similarly, all insurgents are not motivated by grievances. Many insurgents join the insurgency for selfish considerations. For these kinds of insurgents, insurgency is a source of livelihood. Thus, both the theories have some elements of truth and are flawed as well. These two theories have to be supplemented by other theories of violence.

Theory of structural violence [15] and cultural violence [16] developed by Norwegian peace activist Johan Galtung is useful in explaining and understanding the Maoist insurgency. Structural violence has been defined by Galtung as difference between actual and potential. Greater the difference greater the violence and lesser the difference lesser the violence and no difference no violence. This difference between potential and actual is caused by the structures of inequality, illiteracy, unemployment, poverty, malnutrition, deprivation, disease and untimely death. Violence is symptomatic of these structural causes. Unless and until these structural causes are not addressed, Maoist violence/insurgency cannot be addressed. If we apply the theory of structural violence to understand and explain Maoist insurgency, we find wide difference between potential and actual in the case of adivasis, dalits and lower caste people, who constitute the main support base of Maoists.

If we take the different indices of human development index (HDI) such as life expectancy at birth, infant mortality rate, maternal mortality rate, per capita income, literacy, etc. and treat all-India average as the potential and compare it with the actual average of adivasis and dalits—the two most marginalised sections of Indian society and Maoists supporters—we find a wide gulf between the potential and the actual. Below average somatic and mental realisations of dalits and adivasis in comparison to other Indian communities explains why dalits and adivasis extend support to Maoist insurgency.

Theory of cultural violence has been defined by Galtung as justification of violence on cultural grounds such as caste, community, religion, gender, class, etc. Cultural violence aims at transforming the colour of a violent act from red (wrong) into green (right) and if not green then at least yellow (acceptable). In India, violence against dalits and adivasis are justified on cultural grounds such as lower caste and social status. Both the communities are treated as sub-human beings in the hierarchical Indian society. Dalits and adivasis subjected to cultural violence as said earlier, are the main support base of Maoists, whose cause Maoists claim to be espousing. Galtung's concepts of structural and cultural violence can be applied to explain and understand Maoist insurgency.

Theory of relative deprivation [17] developed by Ted Gurr is also useful in explaining Maoist violence in India. Ted Gurr argues that not absolute poverty but relative poverty or social inequality is the real cause of violence. If all are poor then people don't complain but if some amidst them are enormously rich and majority extremely poor and deprived, then this creates fertile ground for eruption of violence and insurgency. In India, the states worst affected by Maoist violence are Bihar, Jharkhand, Odisha, Chhattisgarh, and Telangana, characterised by wide gulf between the poor and the rich. If very few are extremely rich then the majority are impoverished, thereby generating a feeling of relative deprivation, paving the way for rebellion by underprivileged against the privileged. All theories of violence can be applied to Maoist insurgency to explain and understand it.

However, all theories of violence have their limitations too as all of them emphasise on a single cause or factor. Whereas Maoist insurgency is the cumulative result of various factors and causes, organisation and leadership is the most significant. The role of organisation (agency/party) and leadership (actor) in the Maoist insurgency is critical. From the very beginning, the radicals felt the need of organisation and party. They felt disenchanted by the moderates among the Communists, who have taken to parliamentary path and have deviated from the revolutionary path. Instead of challenging and changing the system, these mainstream Communist parties viz. Communist Party of India (CPI) and Communist Party of India (Marxist) have become defenders of the system/state.

The history of Maoist parties can be traced back to Andhra Committee of CPI, which advocated the most radical path against the newly independent Indian state under the leadership of Prime Minister Nehru during the Telangana peasant rebellion against the Nizam of Hyderabad and the then Indian state after the merger of Hyderabad into Indian Union. Andhra Committee was in favour of intensifying the agrarian struggle against the state and spreading it to other areas beyond Telangana to prepare the ground for Communist takeover of India for the emancipation of masses. However, moderate groups of Communists led by S.A. Dange and B.T. Randive argued in favour of giving more time to Nehru to bring agrarian reforms by radical land reforms and zamindari abolition. Dange group, who were most moderates and favourably inclined towards Nehru, later consolidated themselves into present CPI. More radical communists led by

Randive split from parent CPI to form CPI (Marxist). Most radical among communists, the Andhra Committee grew into CPI (ML-Marxist-Leninist), which led the Naxalite movement in late 1960s and early 1970s.

From Proliferation to Polarisation of Maoist Parties

Since its inception, naxal/Maoist parties were characterised by factionalism and split, which was triggered both by ideological differences and personality clashes. However, Maoist parties moved from proliferation to polarisation in late 1990s. Idea of bringing all Maoist parties under one umbrella was mooted long back by Kanai Chaterji of MCC and Kondapalli Sitramaiyah in 1980s too, but it could not make much headway because of serious differences among the top leaders and turf war among various Maoist parties in different parts of India. However, the process of unification of Maoist parties was undertaken seriously in the late 1990s. The lead in these efforts was taken by two leading Maoist parties MCC which was active in Bihar and the PWG in Andhra Pradesh. Smaller Maoist outfits began uniting either with MCC or PWG. The most important breakthrough, however, came in September 2004, when the two most important Maoist parties, MCC and PWG decided to merge into a new party called CPI (Maoist). [18]

The formation of CPI (Maoist) in 2004 was an important milestone in the Maoist movement. Now Maoists are united and well organised. They have also formed an armed wing of the party called Peoples Liberation of Guerrilla Army (PLGA). As per the government estimate, the number of PLGA guerrillas is more than ten thousand. They are equipped with sophisticated weapons snatched from security forces, looted from police armoury and procured from illegal arms bazaar. Apart from PLGA, CPI (Maoist) has also formed scantily armed peoples militia consisting of local sympathisers, who provide logistical support to PLGA in its protracted struggle against the state. Masses living in the Maoists areas provide the intelligence to PLGA to help it in its fight against the state. Guerrillas share cordial relations with the masses. Mao has likened Guerrillas to fish in the water. As fish cannot survive if water becomes hostile, in the same fashion, guerrillas cannot survive if masses become hostile to guerrillas. Continued existence of Maoist guerrillas despite counter insurgency operations launched by the state against Maoists is an ample proof of Maoists enjoying considerable support among the masses.

Role of Organisation/Party in Maoist Insurgency

Since the unification of all significant Maoist parties under one party CPI (Maoist), the left wing extremists are posing potent challenge to the state. The area under Maoist influence has increased rapidly and now one-third of the country is under Maoist influence. This compelled former Prime Minister Manmohan Singh to describe Maoist insurgency as the single largest internal security threat, bigger than terrorism, communalism and separatism. Coming together of all Maoist parties has given a new vigour and vitality to the movement, which is reflected in the increasing assertiveness of Maoist and spiral of violence and counter-violence unleashed by the Maoists and the state.

Politicisation and mobilisation of masses by the Maoist parties is an important factor in the expansion and deepening of the Maoist insurgency. Poverty, inequality, deprivation create fertile grounds for the eruption of LWE but in absence of the organisation/agency/party, these conditions not necessarily lead to insurgency/violence. Bundelkhand region in Uttar Pradesh/Madhya Pradesh is extremely backward but because of the absence of politicisation and mobilisation by a party/organisation, the poor people have developed apathy to their deprivation and resigned to their fate and so inequality, poverty, deprivation of the region has not resulted into any violence. Similar is the case with the famine stricken districts of Kalahandi and Bolangir in Odisha, which are relatively free from Maoist violence despite being the most backward districts of Odisha. Whereas Koraput and Malkangiri are the worst Maoist violence affected districts of Odisha, despite being relatively prosperous in comparison to Kalahandi and Bolangir. Here too, weak presence of Maoist party in Kalahandi and Bolangir is the main reason for relative absence of violence in these two districts despite extreme poverty.

Role of Leadership in Maoist Insurgency

Apart from organisation, the role of leadership is also critical in expanding and deepening the Maoist movement. Imaginative and creative leadership is crucial for the ushering in of successful revolution against an exploitative and oppressive regime. Role of Lenin in Russia, Mao in China, Ho Chi Minh in Vietnam, Chi Guerra and Fidel Castro in Cuba, was critical in bringing successful communist revolution in these countries. In the Indian context, role of Charu Mazumdar was critical in the naxal movement. After his death in 1972, the whole movement collapsed without an effective leadership. Kondapalli Sitaramayh, Kanai Chaterji, Ganpati, Kishen Ji in their own way influenced and shaped the Maoist movement in contemporary period. Death of top leaders is always a great setback for a movement. The recent Lalgarh movement in West Bengal's districts of Midnapur, Purulia and Bankura was practically single-handedly organised and led by top Maoist leader Kishen Ji. Once he was killed by the West Bengal Police in November 2011, the whole movement collapsed like a house of cards. Now Maoists have been weakened considerably in the absence of Kishan Ji's leadership.

However, leadership and organisational factors have only multiplier effects. If other factors such as poverty, inequality, deprivation, etc. are already present only then the leadership and organisation (party) can play a decisive role. Party led by decisive leader will exploit the situation to politicise and mobilise the masses against the system/state and create conditions of insurgency with an eventual objective of capturing the political power. The party and leadership cannot create objective conditions for insurgency. The objective conditions are prior to organisation and leadership. Party and leaders enter into the picture to create subjective conditions for insurgency and successful overthrow of the existing oppressive system.

Geographical Factors

Geographical terrain is another significant factor in insurgency. Difficult terrain characterised by rugged mountains, dense forests, deep ravines, etc. provide ideal environment to insurgents for their hide-outs, training camps and enable them to launch operations against the state represented by the security forces. Maoists intend to create their base in inaccessible areas. Maoist compact revolutionary zone (CRZ) from Nepal borders in north to Telangana-Andhra in south, is a forested and hilly area inhabited by indigenous people (adivasis). Maoists intend to expand their bases from inaccessible areas to gradually engulf the whole of country. It is easier for Maoists to engage their enemy in forested mountain areas to beat them and consolidate their positions. Inaccessible forest areas are bastion of Maoists. In China, Vietnam and other countries too, rebels chose difficult and inaccessible areas to launch their insurgency against the state. Urban areas are called graveyards of guerrillas. Forested and rural areas are Maoist's natural sanctuary. Dandakarayana region stretching from Jharkhand, Odisha, Andhra, Telangana, Chattisgarh, Madhya Pradesh, and Maharashtra is the stronghold of Maoists in India because of the rugged terrain and dense forests.

Koraput and Malkangiri districts of Odisha can be cited again to illustrate the significance of geographical factors in insurgency. Since these two districts are at the tri-junction of Andhra Pradesh, Odisha, Chhattisgarh and falls in Dandkarayana region, these two districts are the worst affected by Maoist insurgency, whereas Kalahandi and Bolangir are relatively less affected by Maoist insurgency despite being extremely poor and backward. This is mainly because of the geographical proximity of Koraput and Malkangiri to the conflict zone in Andhra and Chhattisgarh.

However, geographical contiguity and poverty is not adequate to fuel insurgency. For example, eastern Uttar Pradesh bordering Nepal is free from insurgency whereas territory just beyond the international border in Nepal is firmly under the grip of Nepali Maoists. Despite the porous Indo-Nepal border, Maoists of Nepal have not been able to influence the poor people on the other side of the border in India. This is mainly due to absence of effective organisation (party) and leadership to mobilise the poor and exploited against the state for Maoist revolution.

Conclusions: The Way Forward

There cannot be one size that fits all explanation for Maoist insurgency. Greed and grievance, structural and cultural violence and relative deprivation theories are no doubt significant theories, which offer valuable insight in understanding the Maoist insurgency. But to confine only to the factors/causes of violence given in these different theories and ignore other factors/causes will be an oversimplification of the complex issue of Maoist insurgency. LWE is the result of interplay of several socio-economic and political factors. Every region, every community and every class have their own specific reasons to extend support or oppose Maoism.

To unravel the causes of Maoism in a particular state or region, all relevant factors have to be taken into account by going beyond structural-cultural violence, relative deprivation, and greed-grievance theories. Apart from these factors, the organisation, party, agency, actor/leader, geographical terrain, contiguity and other objective and subjective factors such as ideas, norms, morale, etc. have to be examined to identify the exact causes of Maoist insurgency in the different regions of India.

Once the exact causes of Maoism have been identified, these can be addressed by the state for a lasting solution of Maoist insurgency. Broadly, the causes have been divided into greed (law and order) and grievances (socio-economic). The proponents of greed theory treat LWE as a law and order problem and hence advocate law and order approach as a solution to Maoist insurgency. They advocate enactment of new laws to deal with insurgents, devise innovative strategy to fight Maoist guerrillas like Greyhounds in Andhra, COBRA-CRPF, SOG, etc.), strategic hamleting (cutting off insurgents from masses by relocating civilian population), equipping security forces with sophisticated weapons, counter-insurgency operations (COIN), area domination strategy, clear-hold-build strategy, winning over heart and mind strategy, etc. to decimate Maoist insurgency.

Proponents of grievance theory subscribe to socio-economic measures to address the Maoist insurgency. Since they treat LWE as a socio-economic problem, hence offer socio-economic solutions. They hold triple deficits of development, governance and democracy responsible for the Maoist insurgency and so advocate addressing these three deficits by undertaking an all-round intensive development of the Maoist affected regions, improving delivery system of pro-poor social welfare measures and protecting and respecting the rights of the people. These measures will bolster the confidence of the people in the state and address the grievances of the people against the state. Government has initiated special development programmes like Intensive Area Development (IAD) for Maoist affected districts.

Other socio-economic programmes relate to health and education such as National Rural Health Mission (NRHM), Sarva Siksha Abhiyan (SSA), Integrated Child Development Scheme (ICDS), Mid-Day Meal, etc. Through Right to Information (RTI) and social auditing, the problem of corruption in social-welfare programmes and development works is being addressed and accountability fixed. Forest Right Act (FRA), 2006 and Panchayat Extension Services Act (PESA) have been enacted mainly to redress the long standing grievances of adivasis. Under FRA, adivasis have been given the land rights over the land in which they had their habitats and is under cultivation.

Under PESA, community rights were given to adivasis and their gram sabha. Land Acquisition Act, 2013 was enacted to address the land issue so that land issue cannot be exploited by the insurgents to foment trouble among the affected people as it happened in Nandi Gram, Singur in West Bengal, POSCO and Vedanta land acquisition in Odisha, Tata, Essar in Chhattisgarh, etc. A judicious mix of law and order and socio-economic measures can go a long way in addressing the root causes of LWE. However, these two can only weaken insurgency but not root out it lock, stock and barrel.

Maoism is a political problem and so there cannot be a military solution of LWE. Government must engage left extremists in a serious and sincere dialogue to address the issue of LWE. Both the state and Maoists have to abandon their maximalist position that they have right now adopted. The state should abandon its strategy of decimating Maoist as after all they are its own citizens. They have certain genuine grievances and these grievances must be addressed. Maoists should also abandon their protracted armed struggle against the state. Indian state is not the state of Chiang-Kai-shek of China or Tsarist Russian state. India has the second largest armed forces after China and its military is the third most powerful military in the world after the US and China. To humble the third most powerful state of the world and capture political power through protracted armed struggle seems to be an impossible goal.

World scenario has also changed drastically since the twentieth century, when all these revolutions took place in different parts of the world. The only feasible option seems to be capturing the political power through electoral process—as Latin American countries have achieved through the Bolivarian revolution. Maoist hero Chavez in Venezuela had come to power through elections and not through revolution. Why far away Latin America? In neighbouring Nepal, the Communist Party of Nepal (Maoist) led by Prachand and Baburam Bhattarai came to power through elections only. If Maoists of Latin America and Nepal can come to power through elections, then why not the Indian Maoists?

Before Maoists and state enter into any meaningful dialogue, the state has to lift ban on them and make them a legal entity. How can an illegal organisation enter into dialogue with a state? State should also not insist for Maoists disarming themselves before entering into talks. It is absurd to expect Maoists to commit hara-kiri by disarming themselves before the success of the talks. The only feasible action from both the state and Maoists is to declare cessation of hostilities against each other and enter into dialogue without any precondition from any side. When terror and talk can go together with enemy state Pakistan then why insurgency and talk cannot go together with Maoists, its own fellow citizens? When India can hold talks with enemy Pakistan without any precondition then why insistence on pre-conditions for talks with Maoists? Why this double standard with its own fellow citizens. Discussion will pave the way for some sort of amicable solution, where core concerns of both the state and Maoists are taken care of. If Maoists are so confident of the support of the masses, they should test their popularity in the electoral battles and bring about the desired change as per their belief and ideology. This is the only way out of this vexed problem of Maoist insurgency in India.

Notes and References

1. Weber, Max (2015), "Politics as a Vocation", available at: http://www.sscnet.ucla.edu/polisci/ethos/Weber-vocation.pdf.
2. Nisshit Dholabahi (2014), "Army Option vs. Maoists on Table", *The Telegraph*, October 18, available at: http://www.telegraphindia.com/1141018/jsp/nation/story_18939180.jsp#.VLqfYyuUfwg.
3. Jal, Murzban (2014), "Strangers in the Dark: Neo-liberalism and Maoism in India", *Critique*, Vol. 42, No. 4, pp. 607-629.
4. Shining India was slogan coined by BJP leader Lal Krishan Advani in the wake of 2004 General Election, which BJP lost unexpectedly. Election defeat was interpreted by critics as rejection of neo-liberal policies by the Indian electorates.
5. Suffering India term was coined by critics of BJP to decry its

neo-liberal pro-rich and anti-poor policies of Vajpayee led government at the centre.

6. PM Manmohan Singh (2006), *The Hindustan Times*, April 13, available at: http://www.hindustantimes.com/news-feed/nm9/naxalism-biggest-challenge-pm/article1-86531.aspx.
7. The term commanding heights of the economy was coined by Lenin and was used in Soviet Union in their planning. P.C. Mahalnobis and Jawaharlal Nehru borrowed it from Soviets and was extensively used in India's planned development of public sector enterprises, which was assigned pride of place in the mixed economy.
8. Committee on State Agrarian Relations and Unfinished Task of Land Reforms, Ministry of Rural Development, Government of India, Volume I, Draft Report, New Delhi 2009, available at: http://www.rd.ap.gov.in/IKPLand/MRD_Committee_Report_V_01_Mar_09.pdf.
9. Navlakha, Gautam (2012), "Days and Nights in the Heartland of Rebellion", Penguin India.
10. Collier, Paul and Hoeffler, Anke (2000), "Greed and Grievance in Civil War", Policy Research Working Paper 2355, The World Bank Development Research Group, May, available at: http://www-wds.worldbank.org/servlet/WDSContentServer/WDSP/IB/2000/06/17/000094946_00060205420011/Rendered/PDF/multi_page.pdf.
11. As per data available on South Asia Terrorism Portal obtained through RTI from Ministry of Home Affairs 223 districts in 2008, 208 in 2009, 196 in 2010, 182 in 2011 were affected by Maoist violence, available at: http://www.satp.org/satporgtp/countries/india/maoist/data_sheets/2008-2011.pdf.
12. http://mha.nic.in./sites/upload_files/mha/files/NM-SRE-Scheme_ 160614.pdf.
13. http://www.bjp.org/en/media-resources/press-releases/statement-issued-by-shri-l-k-advani-leader-of-the-opposition-lok-sabha-in-raipur-chhattisgarh-april-25-2006.
14. Singh, Tej Pratap (2013), "From Grievance to Greed: Criminalisation of Maoist Movement in Jharkhand", *The Hindu*, March 3, available at:

http://www.thehindu.com/todays-paper/tp-features/tp-openpage/from-grievance-to-greed/article4470779.ece.

15. Galtung, Johan (1969), “Violence, Peace and Peace Research”, *Journal of Peace Research*, Vol. 6, No.3, pp.167-191.
16. Galtung, Johan (1990), “Cultural Violence”, *Journal of Peace Research*, Vol. 27, No. 3, August, pp. 291-305.
17. Gurr, Ted Robert (1970), “Why Men Rebel”, Princeton University Press.
18. “PW MCCI Merged to Form CPI (Maoist)” (2004), *The Hindu Business Line*, October 15, available at: http://www.thehindubusinessline.com/2004/10/15/stories/2004101 5025 619 00.htm.

2

Understanding Nomads in Contemporary India

Chandrakant Puri

Introduction

Though India is considered to be the world's largest democracy, and an emerging economy in the world; there is still a large number of people living in poverty and marginalized communities getting further marginalized in spite of having a planned development for over six decades. De-notified and nomadic tribes (DNTs) are one such marginalized community in India. According to a rough estimate, they are 11 percent of the total population of India but the exact figure is not known as there has been no census of them after 1931. Administratively, they are scattered across the OBC, SC and ST category. Still there are a few communities which are not covered by any of these three categories. Those who are covered are unable to avail benefits due to various obstacles such as absence of citizenship documents etc. As a result, DNTs are among the most underprivileged and destitute communities facing multiple forms of exclusion. Among the DNTs, nomadic mendicants and entertainer communities have been further neglected when they interface with the highly dynamic contemporary world. Several commissions were appointed by the Central and State Governments to study and recommend policies and programmes to improve the socio-economic status of these communities but all these efforts have hardly achieved the desired results.

In this context, the study was conducted by the author to understand the socio-economic and livelihood related issues of these communities. The focus was also on access to education of the children of nomadic mendicants and entertainer communities in Maharashtra in the context of the Right to Education. The study also identified their struggles for survival including accessing public services, changing livelihood options etc. The team visited around hundred villages in Maharashtra for data collection using a mixed methodology.

This paper highlights some of the major findings of the study based on its objectives and suggests policy guidelines and action strategies for the inclusive development of these communities.

About Nomadic and Denotified Tribes in India

Nomadic tribes (NT) and de-notified tribes (DNT) are the most neglected and discriminated marginalized sections of the Indian Society. They are also one of the socially excluded groups. They were branded as criminal tribes by the Britishers. DNT were traditionally pastoralists, hunter gatherers, traders, transporters, craft workers, entertainers, foragers etc. They are losing their traditional livelihood and are unable to equip themselves with new skills and techniques for survival.

One of the key problems in putting these communities on to India's developmental map is their invisibility due to non-availability of authentic data as post-1931, no caste-wise Census was undertaken in India. Since most of these communities are nomads, they are unlikely to get enumerated in the Census and other data collection process. Hence, they continue to be excluded from the process of development. They also experience atrocities but since they do not have Constitutional protection, they are not covered under the Prevention of Atrocities Act unlike scheduled castes and scheduled tribes. There is also large scale human rights violation against them. A sizable number are denied basic entitlements due to non-availability of basic documents such as land entitlements, ration card, below poverty line (BPL) card, and Aadhar Card (Puri Chandrakant, 2015). [1]

Around 200 communities were notified by the British Government as "criminal tribes" through a notorious piece of legislation called the Criminal Tribes Act, 1871 (Jadhav

Narendra, 2014). [2] The rationale behind the Act was that as per the European view, all gypsies are criminals, and following that logic, all Indian nomadic communities who were gypsies were also believed to be potential criminals. For this reason, there is a large overlap between communities that were declared criminal tribes and those that were nomadic. The Criminal Tribes Act, [3] 1871 identified the following six categories as belonging to the "criminal tribes":

1. Petty traders who used to carry their merchandise on the back of animals and supply to villages, varied items like salt, forest produce etc.
2. Communities that entertained the public through performing arts. Among these were musicians, dancers, singers, storytellers, acrobats, gymnasts, puppeteers and tightrope walkers.
3. Communities that entertained the public with the help of performing animals such as bears, monkeys, snakes, owls, birds, etc.
4. Pastoral groups and the hunting, gathering, shifting cultivator communities within forests that traded not just in forest produce, but in animals as well. They were also herders and traded in meat and milk products.
5. Artisan communities that worked with bamboo, iron, clay etc. and made and repaired a variety of useful articles, implements and artefacts. They traded or sold them to settled villagers.
6. Nomadic individuals who subsisted on charity, or were paid in kind for spiritual services rendered to traditional Indian society such *sadhus, fakirs*, religious mendicants, fortune tellers, genealogists and traditional faith healers; who had a low but legitimate place in the social hierarchy of settled people. Some carried medicinal herbs and provided healing services as well.

Describing the Criminal Tribes Act, 1871 as a blot on the law book of free India, the then Prime Minister Pandit Jawaharlal Nehru annulled it after Independence. As a consequence, these communities were "de-notified", and hence are now referred as "de-notified tribes" (Jadhav, Narendra). [4]

There has been no census of the "de-notified and nomadic tribes" (DNTs) of India after 1931, though their numbers run into millions. They are found in almost all the states, and belong mostly to the OBC category in major states. They are also spread across the SC and ST categories in others. Some communities are not covered by any of the three categories i.e. SC, ST and OBC.

Even those covered under the three categories are often not able to avail of the benefits because of not having caste certificates, or because the quotas are exhausted by the non-nomadic/non-de-notified communities in the reserved categories. Moreover, a number of states have not prepared lists of either the de-notified or nomadic communities, and the status of such people are unknown. As a result, DNTs are among the most underprivileged and destitute communities facing multiple forms of exclusion.

As mentioned above, there is no authentic information available for the DNT communities. Also the National Commission for De-notified, Nomadic and Semi-Nomadic Tribes (NCDNSNT) set up in the year 2006 under the Ministry of Social Justice and Empowerment, Government of India to study various developmental aspects of de-notified, nomadic and semi-nomadic tribes in February 2006 [5] and the Technical Advisory Group appointed by The Ministry of Social Justice and Empowerment, Government of India, on September 8, 2006 [6] failed to give authentic data on the DNT.

Nomadic Mendicats and Entertainer Communities in Maharashtra

As referred earlier, the above two groups were especially found to be vulnerable, hence were focused for the study. As the name suggests, the livelihood of the mendicants is based on the begging in the name of god whereas the entertainer communities entertain people and earn their livelihoods.

Nomadic Mendicants: The traditional occupation of these

communities was preaching, teaching, singing devotional songs and seeking alms. They roamed from place to place leading an ascetic life and due to their religious links, enjoyed a higher place in the caste hierarchy. Hence, most of them are classified in the OBC list in most of the states. As they were landless, most of them have adopted non-agriculture vocations to support them and a large number of them still continue to seek alms. They are found in slums of cities and towns and a large section of them still lead a nomadic life. These communities are smaller in size and dispersed. Each of them claim allegiance to specific deities and have specific place where they return every year on occasions of fairs and festivals. In the rest of the time, they move in small groups of families.

The major communities under this category are: Budbudki, Fakir, Gosavi, Bairagi, Dasari, Garpagri, Jogi, Kashikapdi, Tirumali, Nandiwala etc.

Nomadic Entertainers: The nomadic entertainers earned their livelihood by providing entertainment to the village communities through their acrobatic, jugglery or cultural performances. The demand for their traditional entertainment skills have dwindled under the pressure of modernization. They have been stripped off their artistic dignity and are looked at as begging communities with obsolete skills of entertainment. The names of the major communities under this category are: Mang Garudi, Madari, Nat, Bahurupi, Bal Santoshi, Bharadi, Davri Gosavi, Chitrakathi, Garudi, Ghisadi, Gondali, Kolhati, Madari, Vasudev, Bajikar etc.

The traditional entertainers were landless and hence today they are mostly absorbed in unorganized sector as unskilled workers including petty peddlers, rickshaw pullers and other such menial works. Part of them still lead nomadic life practicing their traditional livelihood skills. They are dispersed in small groups and live with other marginalized groups in cities and towns which provide them opportunities of employment. Nomadic entertainer communities are small in size and in few communities, children play a significant role in earning livelihood.

Patterns of Exclusion among the DNTs

The DNT communities today face multiple exclusions. It ranges from denial of basic amenities, lack of assets of survival to non-availability of work at the village level which leads to migration with the family. The migration further excludes children from education thereby creating generations of illiterates. The working conditions are unhygienic and without any proper facility thereby violating all labour laws. It is especially worst for women as they face even the sexual exploitation from the landlords or the owners at the work sites. They are cheated and exploited financially. On the political front, there is no proper representation due to spread of the community in small settlements. There is also lack of leadership which is not able to unite them. They are getting alienated from their own culture and the rich tradition of music, folklore, mythology, and oral traditions are getting wiped out due to the nomadic way of life and also due to impact of electronic media. What one finds is that gross human rights violation is taking place and there is no one to protect the rights of these communities. This has been discussed in the figure on the next page.

Key Findings of the Study [7]

The study has come out with the following major findings:

1. Majority of the mendicant and entertainer communities belong to Hindu religion. Only two castes, namely Faquir and Darweshi belong to the Muslim religion, which are equally poor and backward.
2. Majority of the families are nuclear (56.6 percent), reflecting change in the family structure among the NT communities.
3. More than 68 percent of the respondents own house, but only 20 percent own agricultural land.

4. Around 40 percent respondents felt their community concerns did not get reflected in the gram sabha. Majority (42 percent) resolve their disputes in jaat panchayat, 20 percent in village panchayat, while only 7.7 percent approach a police station.

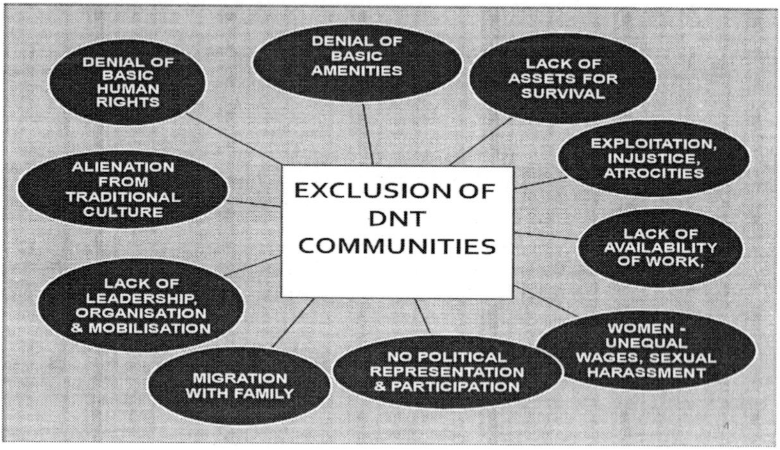

5. Only 10 percent have said that they benefited from the Government schemes.
6. Around 37 percent respondents follow their traditional livelihoods. Majority of those who follow traditional livelihood said it does not generate adequate income for the family. Majority respondents have said that the traditional livelihood adversely affects their socio-economic and educational development and they felt that they get excluded mainly because of their traditional livelihood and the stigma associated with the same. Around 77 percent have mentioned that they would like to change their traditional livelihood.
7. More than 35 percent still migrate in search of employment and those who migrate for survival go with the entire family which adversely affects the education of their children.

8. Nearly 76 percent respondents do not know about the RTE Act, though all parents expressed the need for quality education for their children. Mid-day Meal Scheme has made a positive impact on the enrolment of children in the school. A sizable number of children of the DNT communities are still out of school, including 30.8 percent children in the age group of 5 to 14 years. One-third of the enrolled children drop out between class 5 and 7 and between class 8 and 10. The main reasons for drop out are poverty, lack of interest in education and children contributing to earning a livelihood. A significant improvement is seen for girl child education. Around 63.2 percent girls belonging to NT communities are enrolled in schools.
9. All parents expressed the need for good quality education for their children.
10. Contrary to the popular view, 90 percent respondents have got Aadhar card; 88.5 percent have voter's card; 74 percent have domicile certificates; 78 percent have ration cards; 60 percent have caste certificates and 54.5 percent have bank accounts, which means that they are now in a better position to avail government benefits.

People's Expectations

While on the field, several focus group discussions and case studies were done in order to get insights into the lived experiences of people's lives. The people have expressed their expectations as given in the slide on the next page.

Recommendations for the Development of DNTs

1. There is an urgent need to undertake the census of the DNT communities to systematically plan policies and programmes for them. This also includes updating the ethnographic profile of these communities.
2. The majority of the nomadic mendicants and entertainer families have so far been excluded from the development

process and hardly avail any benefits from these programmes. A time bound plan of action and strategies should be worked out by the governments for their development.

3. Lack of livelihood resources is one of the major problems of DNTs. Hence, there is a need to develop livelihood regeneration programme for DNTs on a priority basis.

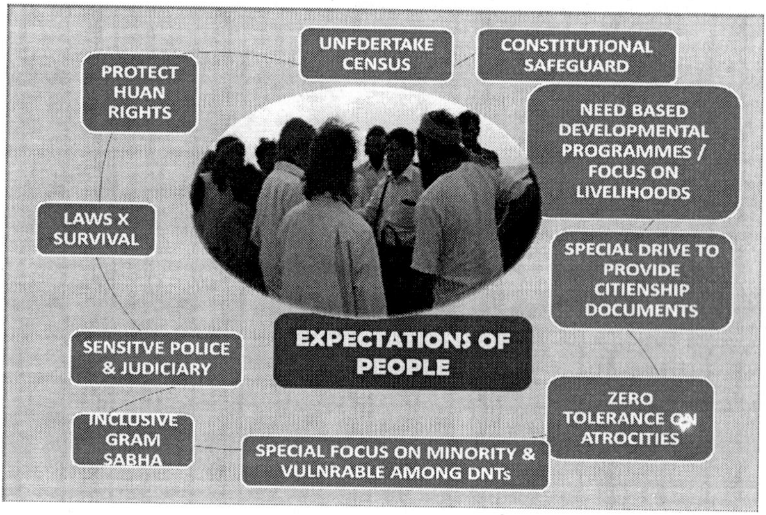

4. NGOs should consider inclusion of DNT in their work and should promote small CBOs/activists working with them to take up the cause of DNTs.
5. Since a sizable number of children of the DNT communities are out of school and face exclusion at various levels, a special strategy needs to be worked out for their inclusion in the education process. The girl child education should be further encouraged by using innovative strategies.
6. There is a pressing need to create awareness on the provisions of RTE Act.
7. School teachers should be sensitized towards the problem of the DNTs and should be instructed to give special

attention for their inclusion in the education system.

Conclusion

This study has been an attempt at creating a major database of the DNT communities whose exclusion results from their historicity and its interface with contemporary processes, deepening their exclusion. The study further locates the disadvantages resulting from such processes including education, changing livelihood pathways and multiple struggles faced by the communities. The relevance of this research lies in building database, new knowledge based on peoples' life experiences and suggestions for reworking the exclusionary patterns of the nomadic mendicants and entertainer communities in Maharashtra for their equitable and sustainable development. This research is an effort to put these communities on to the 'development' map from where they have been largely invisible and start a collaborative dialogue amongst various stakeholders to co-create pathways for transformative change.

Endnotes

1. Puri, Chandrakant (2013), "Evolving Humanity, Emerging Worlds", Manchester, UK; August 5-10, Panel SE-24, "Exclusion of De-notified (Ex-criminal Tribes) and Nomadic Tribes in India: Issues and Challenges for Inclusion", available at: http://www.nomadit.co.uk/iuaes/iuaes2013/panels.php5?PanelID=1563.
2. http://narendrajadhav.blogspot.in (May 31, 2014).
3. Ibid.
4. http://narendrajadhav.blogspot.in.
5. http://en.wikipedia.org/wiki/National_Commission_for_Denotified,_Nomadic_and_Semi-Nomadic_Tribes#Report.
6. http://www.rgics.org/test/wp-content/uploads/2012/09/Technical-Advisory-Group.pdf.
7. Puri, Chandrakant (2015), "Strengthening Right to Education Act Implementation for Children of Nomadic Mendicants and Entertainer Communities in Maharashtra", NEG-Fire, New Delhi and RGCCS, University of Mumbai.

3

Sikh Identity Formation and its Contestations

Kuldip Singh and Sukhjit Singh

The Sikhs are in majority in the Indian state of Punjab. However, at the national level, they constitute an articulate minority whose presence is felt in the socio-economic and political profile of the country. Also at global level, they are a vocal Diaspora having sizeable presence in US, Canada and England. Historically, identity status has been a contested one and still there are elements in the country who consider Sikhs as part of Hindus.

However, Sikhs consider themselves as a distinct community and differentiate themselves from Hindus and others. They often trace evolution of the community from 15th century onwards when Sikh Gurus created a distinct community of their followers by creating unique systems which did not exist before.

This process got institutionalized in the creation of *Khalsa* in1699 with the endowment of five symbols for the Sikhs. The rule of Maharaja Ranjit Singh and subsequently that of British facilitated such an evolution by giving recognition to separateness unleashed earlier.

The real twist to the project of identity formation was discernible in the period of Arya Samaj-Singh Sabha controversy, wherein diatribe against Sikhs was unleashed by Hindu communal organizations depicting the founder of Sikh faith as ignorant and hypocrite, Sikhism as falsehood and Sikhs idolaters. The Hindu ideologues concluded that the Sikhs were Hindus, the claims which were refuted by the Sikhs vehemently.

In the post-Independence period also, Hindu organizations

continued to plead that *Khalsa* was an armed wing created for the protection of Hindus and that Sikhism was a bubble arisen on the Hindu ocean destined to merge. Such propaganda combined with the act of Hindu organizations of disowning Punjabi language, with a view to reduce Sikhs to a minority and preventing the creation of Punjabi Suba, had adverse impact on the Sikhs, similar to the one which Samaj-Sikh confrontation had before Independence.

The armed clash between the followers of radical Sikh preacher Jarnail Singh Bhindrawale and the Nirankaris in Amritsar on April 13, 1978 proved to be a turning point in the history of Punjab politics.

The year is regarded as the beginning of militancy that rocked the state throughout 1980s and early 1990s. With Bhindranwale becoming the rallying point of militant movement, there started the process of radicalization of Akali politics, bringing the issue related to Sikh religion and history to the forefront.

The repressive measures adopted by the government for curbing militancy resulted in large scale violations of human rights which became instrumental in the revival of radical politics by Sikh leaders.

There came a phase when Hindu organizations made Sikh symbols the target and asserted that the community of such symbols was to be driven out to Pakistan. Sikh places of worship were not spared and Golden Temple, supreme religious place of the Sikhs was attacked by the Indian army in 1984, leading to a situation where entire Sikh community lost psychologically. All this gave sustenance to the demand for a separate Sikh state which it lacked.

In such a context, the present paper examines the evolution of Sikh identity from the period of Gurus to the recent phase of militancy in Punjab where identity consciousness assumed the form of a demand for a separate nation. The paper analyses the factors which led to the conversion of identity question of the Sikhs to the nationality question wherein idioms were woven around religion, history and cultures. In the concluding part, it underlines the lessons learnt from the handling of Sikh identity question for managing multi-ethnic societies.

Section I

There were no defined boundaries between the Hindus and Sikhs in terms of religious worship, beliefs and everyday life till the mid-19th century.

In fact, the Sikhs were enumerated as a sect of Hindus in the Census of 1855, the position that was altered in 1871, when Sikhs were enumerated as a separate community (Singh, Sangat, 1999:134 and 136). However, the process of identity formation of the Sikhs is often traced back to Guru Nanak, the first Guru of the Sikhs, who created a distinct community of his followers, different from both Hindus and Muslims, creating institutions which did not exist before (Singh, Gopal, 1994:24).

The martyrdom of the fifth Guru, Guru Arjun Dev by Mughal emperor Jehangir was a watershed event in the history of Sikhs, leading to the conversion of Sikhs from a peaceful community to a militant one and thus creating community feeling among them, in a phenomenal way. The process got institutionalized in the creation of *Khalsa* (pure) on April 13, 1699, which gave a distinct identity to the Sikhs in the form of five symbols: *kes* (uncut hair); *kachha* (a knee length pair of breeches); *kangah* (comb); *kada* (a steel bracelet); and *kirpan* (a sword).

Ranjit Singh's empire gave unity to diverse and scattered elements and moulded the Sikh nation into a well ordered state after tenth Guru had developed a sect into a people and had given application and purpose to institutions created by the first Guru (Cunningham, 1966:120).

It was held that the customs and language of the Sikhs was like that of Hindus but in religious faith and worldly aspirations, they were wholly different from other Indians (Cunningham, 1966:75-76).

From 1873 onward, when Singh Sabha Amritsar was formed, with the objective of restoring pristine purity of Sikhism and bringing the apostates back into the Sikh fold, search for identity and self-assertion became distinctly visible (Singh, Harbans, 1984:23).

Such an urge was heightened by the erosion of Sikh identity during the rule of Maharaja Ranjit Singh which witnessed the influx of a large number of people to the Sikh faith, not because of their allegiance to its principles and traditions but for material considerations (Singh, Harbans, 1984:26).

In fact, the last days of Maharaja Ranjit Singh saw the rise of Dogras and Brahmans, who under the cloak of Hindu-Sikh unity put idols in the Sikh shrines 'overtaking the monotheism of the Sikh Gurus' (Singh, Sangat, 1999:133). The Udasis, who exercised control over Sikh shrines, allowed such activities to take place because they regarded Sikhism as no different from Hinduism in its social milieu (Singh, Sangat, 1999:133).

Following the demise of Sikh rule, Maharaja Dalip Singh's embracing of Christianity in 1853 paved the way for proselytizing activities of Christian missionaries. Under such circumstances, the act of four Sikh pupils of the Amritsar Mission School renouncing Sikh faith in favour of Christianity in 1873 created flutter in Sikh circles.

Therefore, the Singh Sabha movement set out the task of articulating, as a Sikh scholar has put, the "inner urge of Sikhism for reform and gave it a decisive direction, quickening its latent sources of energy. It rescued the Sikhs, awakening in them a new awareness of their past and of the excellence of their faith" (Singh, Harbans, 1984:33).

The advent of Arya Samaj in Punjab in 1877 emerged as an important catalyst for strengthening identity consciousness among the Sikhs. In the early days of reform, the Arya Samaj and Singh Sabha had cordial relationship, as was evident from the tradition of opening up the holy book of the Sikhs Guru

Granth Sahib in the Samaj meetings and from the act of many prominent Sikhs working in tandem with the promoters of the Arya Samaj movement (Singh, Teja, 1984:39).

The era of Arya Samaj-Singh Sabha antagonism started with Arya Samajists attacking Sikh Gurus, which began with the derogatory language used for Guru Nanak and his followers by Swami Dayanand in his book *Satyarth Prakash* (1874), wherein Guru was called *Dhurta* which meant rogue, cheat, fraudulent, crafty, cunning, dishonest and mischievous and also the hymns of the Guru Granth Sahib were described as *mitya* (falsehood) and Sikhism a *jal* (snare) for robbing and cheating simple folk of their wealth and prosperity (Singh, Ganda, 1977:326).

In the second edition, Guru Nanak was depicted as illiterate, self-conceited and hypocrite and the Sikhs were dubbed as arrogant and slaves to lust (Singh, Ganda, 1977:326). Dayanand during his visit to Amritsar in 1877 is reported to have said, "Sikhism was one of the innumerable cults of Hinduism...Nanakji had noble aims but he had no learning" (Jones, Kenneth, 1977:332).

He depicted Guru as the one knowing the language of the villagers but having no knowledge of Vedic scriptures or Sanskrit. The followers of Guru despite the fact that they did not worship idols were accused of being idolaters because worshipping Guru Granth Sahib was labelled as idolatry.

At the Lahore session of Arya Samaj held on November 24-25, 1888, Pandit Lekh Ram, an important leader of Arya Samaj placed the holy book of the Sikhs, ''Guru Granth Sahib before him on a table in a manner which was taken as sacrilege by the Sikh leaders (Singh, Ganda, 1977:328). Lala Guru Datt, another influential leader of Arya Samaj, claimed that Guru Gobind Singh was not even a hundredth part of Maharishi Swami Dayanand Sarswati and it was difficult to say whether the Sikhs had any religion (Jones, 1977:336).

Such a competitive communal politics led Sikhs to break relations with the Samaj and join Lahore Singh Sabha. The tussle between the two communities over religious issues and their sense of identity, reached feverish pitch, in the aftermath of these incidents (Jones, 1977:330).

In 1898, the issue of Sikh identity took a new turn, when following the death of Sardar Dyal Singh Majithia, who had left vast wealth in the name of Dyal Singh Trust, his widow contested the ownership of the property claiming that Hindu law of inheritance invoked for giving property to the trust did not apply because her husband was a Sikh and not a Hindu. In the ensuing litigation, the Punjab High Court ruled that Sardar Dyal Singh Majithia was a Hindu (Jones, 1977: 344).

This set off an intense debate over the status of the Sikhs to which Lala Thakar Dass and Bawa Naryan Singh in their books with identical titles *Sikh Hindu Hain* in 1899 declared that the Sikhs were Hindus (Jones, 1977:345). To this, Bhai Kahan Singh Nabha, a Sikh ideologue, in his book *Hum Hindu Nahin* (1899), which is widely accepted as Sikh view on identity question, responded arguing that the Sikhs were not Hindus (references to Nabha's book are from, Grewal, 1999:231-251). Nabha claimed that the *Khalsa Panth* was distinct from both Hindus and Muslims and those who refused to accept this were ignorant of Sikh scriptures and history.

To the view that the Sikhs were Hindus because they had emerged from Hindus, he argued that Christianity having emerged from Judaism and Islam were regarded as distinct religions. Sikh scholars while following Nabha, stressed that if most of the Sikhs were converts from Hindus, so were Indian Muslims and Christians, who cease to be part of Hindu way of life after conversion to the new religion (Narang, 1983:117).

To the claim that the Khalsa Panth was meant for spreading Hindu religion and that Guru Tegh Bahadur, the ninth Guru of the Sikhs, sacrificed his life for the sake of Hindus because the Guru himself was a Hindu, Bhai Kahan Singh argued that Guru Tegh Bahadur sacrificed his life, not for a sect but for upholding the principle, i.e protecting the oppressed ones. He justified Sikh claim for separate identity as they had their own scripture in Guru Granth Sahib, conception of *karma* (destiny), *upasna* (worship) *and gian* (spiritual knowledge). The notion of equality, discarding caste based distinctions, denouncing of Hindu social order, by Guru Nanak and his successors and adoption of prescribed Sikh symbols, gave Sikhs a distinct identity.

Towards the turn of 20th century Sikhs lost control of their places of worship to non-Sikhs and there reached a stage when the very existence of their faith was in danger (Akbar, 1985: 94). The Sikhs were returning to Hinduism and the fear of re-absorption was heightened with proselytizers insisting that Sikhs were only Hindus by another name (Akbar, 1985:94). As a result, distrust and conflict became the hallmarks of Punjabi society.

As Kenneth W. Jones has written, "pattern of conflict became institutionalized. Provocations produced set responses. What has been implicit in the nature of Punjab society became explicit. Tensions might increase or decrease but beneath the surface, fears, suspicion and hatred remained" (Jones, Kenneth, 1977:42-43). All this strengthened Sikh conviction, inculcated by Singh Sabha Movement, that they were a separate community and their resolve for securing separate rights and institutions for protecting Sikh interests was further bolstered (Singh, Teja, 1984:43).

As such, the Sikhs started emphasizing that "Guru Nanak was not to be seen as merely another Hindu reformer but as a self-conscious founder of the new religion" (Brass, 1974:282).

The attempts to portray Sikhs as part of Hindus by Hindu organizations continued even at the dawn of the 20th century. In June 1900, Swami Shardanand of Arya Samaj performed *shudhi* (purification) ceremony of Rahtia Sikhs, who were weavers by profession, giving them Hindu sacred thread after shaving of their hair and beard, which led to outrage among the Sikhs (Jones, 1977:348-349).

The Sikhs alarmed at the joining of Arya Samaj by lower castes Sikhs like Mazbhis, Rahtias and Ramdasis, formed the

Khalsa Biradari in 1908 with a view to modify Sikh attitudes toward untouchables (Barrier, 2000:76). The Chief Khalsa Diwan was founded in 1902 for the welfare of Khalsa, preaching Gurbani and safeguarding the political rights of the Sikhs.

In 1905, it succeeded in getting the Hindu idols removed from the Golden Temple, which was not taken kindly by Hindus and in 1909 with its efforts, Anand Marriage Bill was passed, giving legitimacy to marriage ceremony exclusive to the Sikhs (Barrier, 2000:77).

Following the Lucknow Pact, signed in 1916, wherein Congress conceded 50 percent representation to the Muslims in Punjab; Sikh leaders started demanding separate electorate for the community, which Punjab Provincial Congress Committee, dominated by Arya Samajists did not endorse, contending that Sikhs being part of Hindus were not entitled for separate electorate (Singh, Sangat, 1999:158).

In 1921, Raja Narendra Nath of Hindu Mahasabha, in the Punjab Legislative Council held, "I look upon Sikhs as higher Hinduism. The Granth Sahib is nothing more nor less than the higher teachings of the Vedas and Upanishads" (Akbar, 1985:134). To this, Sikh leader Mehtab Singh retorted, "...if the Sikhs do not wish to remain in the fold of Hinduism, why should the Hindus seek to force them to do so...? We wish to manage our own affairs and look after our Gurudwaras and are determined to do so" (Akbar, 1985:134).

The process of social differentiation between the Sikhs and Hindus got sharpened as a result of the enactment of the British Act of 1921, creating separate electorates for the Sikhs (Brass, 1974:319).

The Shiromani Gurudwara Prabandhak Committee (SGPC) created in November 1920, for managing Sikh shrines and Akali Dal a political party of the Sikhs, coming into being in December 1920, spearheaded the Gurudwara reform movement which on the one hand contributed to the cause of national liberation because the control of British supported

Mahants (priests) on the Sikh shrines was put to an end and it on the other hand, "consolidated and politicized Sikh identity and community consciousness" (Singh, Gopal, 1994:50).

The struggle resulted in the passing of Sikh Gurudwara Act, 1925, which sharpened the consciousness of Sikh separateness (Brass, 1974:283). Hindu priests were removed from Gurudwaras, even Hindu idols were broken, to which Hindus reacted very sharply and it resulted in social distancing between the two communities (Nayar, 1966:67).

The demand for communal representation was intensified further, promoting a sense of self-consciousness and identity among the Sikhs (Brass, 1974:309). In 1928, Sardar Sunder Singh Majithia, while presenting the case of the Sikhs before the Indian Statutory Commission, on behalf of Chief Khalsa Diwan, remarked, "From the time of the Gurus, Sikhs have been a distinct community and I cannot accept the statement from a rival community that we are not a distinct community…Religiously and socially we are a distinct community and as such our interests are not identical with those of any other community at all" (Brass, 1974:284).

In December 1947, Gandhi held that Sikhism along with Buddhism and Jainism was the offshoot of Hinduism (Singh Sangat, 1999:259). Such assertions helped to keep the urge for separate identity among Sikhs alive and even gave Akali Dal, justification for waging struggle for the creation of Punjabi Suba, a province within the Indian union, based on Punjabi language in Gurmukhi script, thus bringing Sikh identity to the forefront and in the process, pushing Punjabi identity to background (Brass, 1974:323). Hindus disowning Punjabi in favour of Hindi as their mother tongue in 1951 and 1961 Census gave impetus to such a demand.

The act of disowning of Punjabi was perceived by the Sikhs as a move to reduce them to linguistic minority besides being a religious minority and there arose "two different but parallel streams of cultural nationalism, both anchored around the religious identity of their respective communities" (Nanda,

1988:13).

The creation of language-based province was considered essential for preserving the Sikh culture and protecting Sikh religion. As Sikh leader Master Tara Singh said in 1955, "I feel we cannot save our religion without attainment of political power in the Punjabi speaking region...For the sake of religion, for the sake of culture, for the sake of panth and to keep high the flag of Guru, the Sikhs have girded their loins to achieve independence" and that his task was to "free Sikhs from Hindu dominance" (Quoted in Nayar, 1966:108 and 36-37). Hindu leaders held that the demand for language-based province was merely a camouflage for attaining Sikh hegemony and eventually Sikh sovereign state.

This led them to disown Punjabi in favour of Hindi as the mother tongue, as the purpose was to show Punjabi speaking people in minority and demolish the claim for language-based state (Nayar, 1966:42 and 44).

Arya Samajists even held that maintaining of Sikhism as a separate religion was not required because the objective of protecting Hindus against Muslims, for which Khalsa was created by the Guru, was no longer valid as the Muslims were evicted to Pakistan and therefore, Sikhs should revert back to Hindu fold (Sarhadi, 1970:160).

The Sikh leader Master Tara Singh asserted that Sikh culture was different from Hindus. Therefore, Sikhs wanted to have a province where they could safeguard their culture and tradition (Rai, 1965:273). To sabotage Akali demand, communal Hindu organizations demanded the formation of Maha Punjab comprising Punjab, PEPSU, Himachal Pradesh and parts of Western U.P., apparently with a view to keep the Sikhs in minority (Rai, 1965:287-288).

In its report submitted in 1955, State Reorganization Commission (SRC) did not concede Sikh demand following which the political climate in the state became communally surcharged. Gyan Singh Rarewala, ex-Chief Minister of PEPSU, called the Report a death-knell for the Sikh community. Master

Tara Singh declared it 'a decree of Sikh annihilation' and opined, "We are face to face with the calamity greater than that of 1947. The catastrophe of 1947 finished thousands of Sikhs. But the report of SRC wipes us from the face of the world" (Akbar, 1985:151).

Hukam Singh blamed Nehru and called him the "spearhead of militant Hindu chauvinism, who glibly talks about nationalism". The Hindu slogan *Hindi, Hindu and Hindustan* got prompt reply from the Sikhs with *Dhoti, Topi Jamuna Par* that is to send the wearer of dhoti and cap across river Jamuna (Akbar, 1985:150).

Section II

Finally, language-based province was carved out in 1966 bifurcating the old one, demarcating Punjabi speaking areas for Punjab and Hindi speaking for Haryana. Even after the creation of the new province, Akali Dal could not secure political power on regular basis, which many believed to be the real motive behind this Akali demand.

Therefore, search for effective agenda that could secure political power on more regular basis continued. The adoption of Anandpur Sahib Resolution in 1973, projected by Akali Dal as an agenda for securing more state autonomy and a truly federal structure in India, was interpreted as a desperate attempt on the part of Akalis for establishing enduring support base among Sikhs. The Congress and other political groups opposed to Akalis considered the resolution passed by Akali Dal as sectarian and secessionist.

Thus, the Anandpur Sahib Resolution acquired an exclusive Akali identity. President of Akali Dal, Harchand Singh Longowal at a later stage declared that the prime cause behind the passing of the resolution was the apprehension that like Buddhism and Jainism, the Sikhs might lose their identity to the vast ocean of overwhelming Hindu majority (cited in Akbar, 1985:173).

Moreover, Sikh ideologue Bhai Kapur Singh's interpretation of the litany *Raj Karega Khalsa* to mean that the

Sikhs are *Raj Jati* destined to rule and that the Anandpur Sahib Resolution only reasserted the political goal of the Sikhs set out by the Tenth Guru of the Sikhs, that aspired the pre-eminence of *Khalsa,* served as a permanent factor for distancing Sikhs from the Hindu community (Bombwall, 1987:159).

As noted earlier, the armed clash between the followers of Sikh preacher Jarnail Singh Bhindranwale and the Nirankaris in Amritsar on April 13, 1978 proved to be a turning point in the history of Punjab politics. This proved to be the beginning of turmoil that rocked the state throughout 1980s and the early 1990s. The murder of Nirankari chief in April 1980 brought radicals among Sikhs to the forefront.

The assassination of Lala Jagat Narain in August 1981 for which Bhindranwale was the main accused, turned out to be a major event in Sikh politics. His surrender to the police on September 20, 1981, in the presence of his followers, at Gurudwara Gurdarshan Parkash at Mehta Chowk near Amritsar, brought him to the centre stage of Sikh politics.

In August 1982, the Akalis joined Bhindranwale for launching *Dharm Yudh Morcha* (Holy War), terminating their Nehar Roko Morcha, launched in1981, on the river water issue. Thereafter, started the process of radicalization of Akali politics, bringing issues related to Sikh religion and history to the forefront.

The adverse reaction of communal Hindu organizations and strict measures adopted by the government for curbing militancy, often led to the allegations of human rights violations, thereby becoming instrumental in the revival of radical politics by Sikh leaders.

Bhindranwale after creating political space, in the agitation launched by Akalis, started appealing his followers in highly subjective idioms, invoking Sikh honour, which in his perception was denied to the Sikhs. With a view to revive Sikh militant traditions, he even declared that, "whosoever insults the Guru Granth Sahib (Sikh holy book) should be killed there and then" (Pettigrew, 1987:16).

The "Sikhs were second class citizens and slaves in India and that they must fight for their independence", he asserted repeatedly (Dang, 1987:238). Some happenings of that period that hurt Sikh sentiments, gave sustenance to his appeal. In 1981, following All India Sikh Students Federation (AISSF)'s demand for putting ban on smoking in Amritsar and declaring it the holy city, Hindu organizations brought out a procession against anti-smoking campaign of the Federation. They shouted provocative slogans like *'Kachha, Kanga te Kirpan, Sab ko bhejo Pakistan'* that is Sikhs wearing such symbols, considered sacred by them, will be driven out to Pakistan.

Subsequently in 1982, during the Asian Games, Sikhs on their way to Delhi were humiliated in Haryana under Bhajan Lal, the then Chief Minister's plan to 'teach Sikhs a lesson' (Dang, 1987:238).

All this made Sikhs feel as a minority, threatened by their own state. In February 1984, the replica of Golden Temple and picture of Guru Ram Das, founder of the city, installed at railway station in Amritsar were smashed, Sikh women were humiliated and sacrilege of Sikh shrines was done in Haryana. After these events, Bhindranwale assumed the mantle of saviour of Sikh faith, which he repeatedly asserted was the target of Hindu majority.

There reached a stage of identity issue engrossing Sikh mind at unprecedented scale. The sense of grievance among them increased several fold, after the army attack on Golden Temple.

While drawing a parallel between the Jallianwala Bagh massacre and the attack on Golden Temple, a noted Sikh historian, Khushwant Singh wrote that, "while the Jallianwala massacre became the turning point in the history of India's freedom movement, the massacre in the Harmandir may well become a similar turning point in the history of the Khalistan movement" (Singh, Khushwant, 1984).

He went to claim that, "Now it will be increasingly

difficult to find a self-respecting Sikh who will have a kind word to say for the Government...The number of innocent people slain will go on being increased, the desecration of the Parikarma and the Akal Takht will be magnified to leave a permanent dent in the bruised memory of the Khalsa Panth...Sikhs will feel isolated and a separate people" (Singh, Khushwant, 1984). He declared that after attack on the Golden Temple, Sikhs felt "isolated and a separate people".

After the killings of Sikhs in Delhi and elsewhere in the country in November 1984, "without exception the entire community lost psychologically—its collective psyche was wounded and its sense of security was lost" (Mathew, 1985: 260).

The involvement of the state in teaching Sikhs a lesson made the situation grim, as overnight the Sikhs felt they had become second-class citizens. In fact, "never before had the Sikhs felt like a minority as after these events" (Puri, 1988:9). The wounded Sikh psyche produced emotional outbursts, alleging that, "Sikhs are slaves in India. Our aim is to break the shackles of this slavery" (AISSF, 1985:1) and that the "poseur, deceptive, liar and cheater rulers of India are scared of the sacred Gurbani of the Sikhs and their dress".

Further, that the target of Hindu communalists and the government, "outrage has been the religious places of the Sikhs and the holy book *Shri Guru Granth Sahib*" and that the solitary aim of the government is to destroy the Sikh tenets and their proud inheritance" (AISSF, 1985:1).

The *Gurmata* (resolution) passed at *Sarbat Khalsa* (Sikh congregation for deciding socio-political affairs of the community), held on Diwali festival on November 1, 1986 also reiterated that the "Hindu Government was continuing its horrendous attack on Sikh tenets and Punjab" and that the Government wanted to "crush the sacred-most and rich religious inheritance of the Sikhs" (*Gurmata,* 1986).

The alienation of the Sikhs, following widespread feeling in the country about the Sikhs in general, as an irrational, sectarian and bloody-minded people, extreme communalists out to destroy the fabric of Indian unity, became a well-recognized fact of the Indian polity (Alam, 1987:82). This resulted in a feeling of distrust in the minds of the Sikhs, the distrust that also emanated from the reservations of communal Hindus in accepting them as a separate community (Nanda, 1988:13).

The grouses Sikhs entertained against communal Hindus and the rulers of the time, reinforced community feeling among them, the status, which they claimed, was bestowed upon them since the creation of the Khalsa.

During the phase of militancy in Punjab, the militant Sikh organizations invoked martyrdom as a symbol for constructing Sikh identity. The analysis of Sikh literature and political writings, produced during this period, brings out that the reviving of such Sikh images was sought, in which Sikh identity was "constructed primarily through the metaphor of martyrdom" (Das and Bajwa, 1994:247).

The punishments that the defiant Sikhs are said to have received at the hands of Mughal emperors because of their protest against injustice, was referred to frequently in the Sikh discourse of the time (Das and Bajwa, 1994:247).

The objective of militant assertion was to "create a homogeneous Sikh community that would resemble a congregation" and the violence resorted for political purposes, "was largely connected with the way the politics of identity influenced the behaviour of the militants to create a homogenous community and anybody opposing the project was the enemy" (Judge, 2005:96 and 98).

They could not succeed in creating homogeneous Sikh community of their mould because "were it possible to build nations through human agency, surely the world would have been full of them" (Brar, 2002:17).

The role of human agency for building ethnic consciousness has its limitations. The Sikh symbols and traditions were revived with vengeance but this happened largely in response to communal and partisan response of Hindu communal organizations and repressive and at times, communal state response.

Section III

The Sikh identity evolved through the period of Sikh Gurus, the process that got institutionalized in the creation of *Khalsa*, which gave distinct identity to the Sikhs in the form of five symbols. During the Sikh rule of Ranjit Singh and British era, the recognition of separate identity strengthened the awareness of distinct identity.

The real twist to identity formation happened in the period of Arya Samaj-Singh Saba controversy, where in the diatribe against Sikhs was unleashed depicting the founder of Sikh faith as ignorant and hypocrite, Sikhism as falsehood and Sikhs idolaters and the Hindu ideologues even claiming that the Sikhs were Hindus.

In response to such accusations and with fear of being reabsorbed in Hinduism, the Sikhs started distancing themselves from Hindus. In post-Independent period also, the argument that the Khalsa was an armed wing meant for the protection of Hindus and that Sikhism was a bubble arisen on the Hindu ocean, destined to merge, combined with the role of Hindu communal organizations in the disowning of the Punjabi, were not taken kindly by the Sikhs.

The process of the evolution of Sikh identity brings out that denial of identity status and the fear of being absorbed by hostile community leads to assertions of identity with vengeance.

It also leads to the inventing of beliefs, symbols and practices vital to the existence of community which over the period of time may get moderated and may go the background. If the contestations of Sikh identity status by the hostile community are any guide, it may be underlined that freedom to minorities for professing their faith and asserting their community status is the appropriate mould for peaceful coexistence of diverse communities in modern day diverse societies.

The homogenization drive may prove to be counterproductive resulting in revival of hard line in not only in terms of religious beliefs but might also lead to political demands which at times may assume secessionist undertones.

References

Akbar, M.J. (1985), "India: The Seize Within", UBS Publishers, New Delhi.

Alam, Javed (1987), "Political Implications of Economic Contradictions in Punjab," Gopal Singh (ed.), 'Punjab Today', Intellectual Publishing House, New Delhi.

All India Sikh Students Federation (AISSF) (1985), Resolution Passed at All India Sikh Students Meeting (in Punjabi), Giani Press, Amritsar.

Barrier, N. Gerald (2000), "The Singh Sabha Movement 1875-1925", Harish K. Puri and Paramjit S. Judge (eds.), 'Social and Political Movements', Rawat Publications, Jaipur.

Bombwall, K.R. (1987), "Sikh Identity and Federal Polity: A Critique of Akali Position", Gopal Singh (ed.), 'Punjab Today', Intellectual Publishing House, New Delhi.

Brar, Bhupinder (2002), "One More Mirage: 'Pluralist Nation-Building' in Multi-Ethnic Societies", Gurnam Singh (ed.), 'Ethno-Nationalism and the Emerging World (Dis)order, Kanishka Publishers, New Delhi.

Brass, Paul R. (1974), "Language, Religion and Politics in North India", Vikas, New Delhi.

Cunningham, Joseph Davey (1966), "A History of the Sikhs: From the Origin of the Nation to the Battles of the Sutlej", S. Chand, New Delhi.

Dang, Satya Pal (1987), "Hindu Communal Organisations in Punjab," Gopal Singh (ed.), 'Punjab Today', Intellectual Publishing House, New Delhi.

Das, Veena and Bajwa, R.S. (1994), "Community and Violence in Contemporary Punjab", Denis Vidal (ed.), 'Violences and Non-Violences in India', School of Sciences Studies, Paris.

George, Mathew (1985), "The Victims", Amrik Singh (ed.), 'Punjab in Indian Politics: Issues and Trends', Ajanta Publications, New

Delhi.

Grewal, J.S. (1999), "Nabha's Hum Hindu Nahin: A Declaration of Sikh Ethnicity", Pashaura Singh and N. Gerald Barrier (eds.), 'Sikh Identity: Continuity and Change', Manohar Publications, New Delhi.

Gurmata (1986), Resolution Passed by the Assembly of the Sikhs at Sarbat Khalsa (Sikh congregation) meeting held in Amritsar on November 1.

Jones, Kenneth W., (1977) "Hum Hindu Nahin: The Arya Sikh Relations, 1877-1905", *The Punjab Past and Present*, Vol. 11-12, October.

Joyce, Pettigrew (1987), "In Search of a New Kingdom of Lahore", *Pacific Affairs,* Vol. 60, No. 1.

Judge, Paramjit S. (2005), "Religion, Identity and Nationhood: The Sikh Militant Movement", Rawat Publications, Jaipur.

Nanda, Subrat K. (1988), "Punjab: Case of Fractured Nationality", *Mainstream,* Vol. 26, No. 12.

Narang, A.S. (1983), "Storm Over the Sutlej: The Akali Politics", Gitanjali Publishing House, New Delhi.

Narang, Surjit Singh (2002), "Ethnic Conflict in a Majoritarian Mould: Case of Sikhs in Punjab", Gurnam Singh (ed.), 'Ethno-Nationalism and the Emerging World (Dis)order', Kanishka Publishers, New Delhi.

Nayar, Baldev Raj (1966), "Minority Politics in the Punjab", Princeton University Press, New Jersey.

Puri, Harish, K. (1988), "Akali Dal and National Question", *Mainstream,* Vol. 16, No. 21.

Rai, Sattya M. (1965), "Partition of the Punjab", Asia Publishing House, Bombay.

Singh, Teja (1984), "The Singh Sabha Movement", Ganda Singh (ed.), 'Singh Sabha and Other Socio-Religious Movements in PB', Punjabi University, Patiala.

Singh, Ganda (1977), "The Origin of the Hindu-Sikh Tension in the Punjab", *The Punjab Past and Present*, Vol. 11-12, October.

Singh, Gopal (1994), "Politics of Sikh Homeland", Ajanta, New Delhi.

Singh, Harbans, (1984) "Origin of the Singh Sabha", Ganda Singh (ed.), 'Singh Sabha and Other Socio-Religious Movements in PB', Punjabi University, Patiala.

Singh, Khuswant (1984), "Tears in a Pool of Nectar", *Indian*

Express, June 17.
Singh, Sangat (1999), "The Sikhs in History", UN Common Book, New Delhi.

4

Nationalism in Contemporary India

K.C. Sreekumar

Nationalism occupies the centre stage of intellectual discussion in India today. Both the print and visual media celebrate narratives on nationalism often distorting the very meaning of the concept. While some resort to the classical concept of nationalism to substantiate their arguments on contemporary Indian nationalism, others equate it with democracy, secularism, majority rule, protection of minorities, religious fundamentalism, etc. This necessitates a scholarly pursuit, with the assistance of a theoretical perspective, in understanding not only nationalism but also sub-nationalism in contemporary India. How contemporary Indian nationalism can be explained? What is its true nature? Is it in any way different from the nationalism preached and practiced during the freedom struggle? Is there a pan-Indian nationalism? Or, is it a collection of many sub-nationalisms or mini-nationalisms? Is there an all-India nationalism co-existing with sub-nationalisms? Is sub-nationalism a divisive force?

Can the present day Indian nationalism be termed as Hindu nationalism? What is the force that keeps India united despite the multi cultural nature of its population and the innumerable contradictions inherent in the society? If one has to unravel the nature of contemporary Indian nationalism, he/she has to answer these and perhaps more questions. That is precisely what I would venture to do in this work.

Difference in Perspectives

Nationalism is study from different perspectives by different writers. It has also to be considered that quite often the intellectual upbringing of the scholar, colours his perspective on a particular issue or subject of study. A glimpse of the views of some eminent writers would enable one to see that there is considerable disagreement among social scientists as to what this phenomenon is and how it should be studied. Historians like R.W. Seton-Watson, Carlton J.H. Hayes, Boyd Shafer, Hans Kohn and Hugh Seton-Watson described how the new sentiment of nationalism dissolved the old society with its kings, clerics and provincial loyalties and then provided a common allegiance that gave the reconstructed nations an extraordinary cohesiveness. [1] For Hayes, nationalism is a state of mind akin to religious belief, "an emotional loyalty to the idea or the fact of the national state, a loyalty so intensely emotional that it motivates all sorts of people...". [2] His belief that nationalism was a matter of faith rather than reason made it difficult to explain the phenomenon.

The academic Marxists consider nationalism as an integral part of the ideology of the rising bourgeoisie in pursuit of its class interests. But it is observed that a bourgeois class in a fully fledged condition is rarely found in the Third World countries at the time of independence. It is because the colonial policy, which is cantered upon the interests of metropolitan and international capital, prevents the formation of a national bourgeoisie. [3]

The modernization theory tries to establish a positive connection between nationalism and the process of modernization. In the opinion of Ernest Gellner, the furnace of modernization must melt every society that falls into it, and recast it in a standard mould. He considers the nation as the present standard mould. Every society that gets modernized is bound to develop nationalism and change itself into the standard mould. Nations may differ from one another, but the modernizing process permits each one to have one culture and language. The process of determining what language and culture are more suitable for the purpose of modernization leads to nationalism. [4] Gellner is giving a great deal of emphasis on culture, especially on language. Culture and language are important factors that may help the creation of nationalism. However, Anthony Smith feels that these should not be considered unavoidable for the creation of nationalism. Language might have been important to early nationalists but it was only after they became nationalists that they used language as a unifying device or imposed proper language on the masses after they achieved the power to do so. [5]

The social communication theory and the erosion of primordial loyalty theory are two other theories that try to unravel the phenomenon of nationalism. Karl Deutsch is the chief exponent of the social communication approach. He argues that nationalism is the product of a new type of education which affects the intelligentsia and then spreads to other groups using mass media and the spread of literacy to reach the common people. According to him, what is important is not the presence or absence of any single factor, but merely the presence of adequate communication facilities with enough complementarities to produce the overall result. [6] This leads to rapid social mobilization which encourages both nationalism and sub-nationalism.

This theory is criticized on the grounds that it is irrational to assume that close contact between people would always ultimately lead to nationalism. The theory also leads one to the conclusion that the group with the highest level of social communication is the most nationalistic. One should therefore naturally expect the level of nationalism in the upper classes to be greater than that in the poor and middle class groups. But Kautsky and others do not agree with this proposition. [7]

The erosion of primordial loyalty theory defines nationalism as the existence of shared positive affections towards the nation-state or towards the formation of nation-state. This theory stresses the effect of modernization on traditional patterns of social groupings. Thus, according to scholars like Eisenstadt, Smelser and Silvert, development of nationalism has a positive correlation with the presence of modernization and the related process of structural differentiation. [8] This explains group crystallization at the societal level. But it does not explain why group conflicts occur in a nation-state or why these conflicts do not lead to disintegration of the nation-state.

The fundamental premise of the conflict theories is that competition among groups generates nationalism. These theories explain the development of nationalism in plural societies. They assume that inequality is inherent in all social systems. Increase in the level of inequality, which may be the result of modernization or some other social change generates more social tension which leads to conflict. A conflict situation always intensifies group identification and develops group integration. When the mutually exclusive groups engaged in conflict acquire the aim of the control of the political system of the society, nationalism is generated. Three distinct theories come under the category of conflict theories. They are economic, social-class and ethnic conflict theories. These theories are based on the pluralist mode of society. Albert Breton and Harry Johnson are the two main promulgators of the economic conflict theory which holds that central to nationalism is the conflict over property relationships. [9]

The social-class conflict theory holds that nationalism results from a set of demands advanced by those social classes which have unequal access to the resources allocated by the distributive system. [10] In the ethnic conflict theory, the boundaries of a society are determined by ethnic criteria, and such criteria are considered to be salient sources of contention between the societies. According to Gordon Lightbody, if an ethnic group's demands are more extensive and if these dominate the aggregate number of demands upon a system, that group will come closer to obtaining an autonomous political identity with that territory. [11]

Political anthropology has also made contribution to the study of nationalism. It owes its origin to the studies conducted by social anthropologists in the political behaviour of colonised societies. The cornerstone of this approach is the concept of 'colonial situation'. [12] Nationalism is viewed as an important aspect of the reaction of the 'colonised' societies to the 'colonial situation'. The colonial situation is the setting within which the complex interaction between the colonised societies and the colonisers take place. This interaction was given little attention by social anthropologists prior to Gluckman. [13]

However, it was Georges Balandier who formulated the concept in an advanced manner. In his opinion, indigenous social change can be understood only in an analytical context in which attention is focused on the specific features of the colonial situation. These features are the radical roots of groups, their extreme distinctions, the hostile relationships it involves, and the necessity imposed on them to coexist within the limits of a single political framework. Balandier describes in detail four specific features of the social dynamics created by a colonial situation.

Firstly, in a colonial situation, two distinct and complex cultures or civilizations and two different and dissimilar technologies interact. Out of these two contrasting cultures or civilizations, one is more rationalistic, individualistic, bureaucratic, capitalistic and technologically advanced; and the other more affective, ascriptive, communalistic and technologically unsophisticated. These forces can scarcely cooperate. Their coexistence leads to conflict.

Secondly, two discrete groups, a dominant foreign minority and an indigenous, powerless, minority, articulate the contact between the two cultures. These groups pursue conflicting interests within the same colonial framework.

Thirdly, due to their military power, better organization, tighter central control over modern means of communication and administrative unification of the colony, the foreign minority group is superior to the indigenous majority group. Hence, the inter-relationships between these groups are shaped by the requirements and aspirations of the dominant group or the colonizer.

Lastly, the dominant minority tries to reinforce their material force with a complex fabric of universalistic pseudo justification. Here, one would recall the 'civilising missions' and the 'white man's burden'. The major limitation of this theory is that it deals only with the anti-colonial phase of nationalism and leaves the post-Independence period unexplained. Each one of these theories focuses on one aspect of nationalism only.

Conflict Resolution-cum-Consensus Generation

This discussion makes it clear that the phenomenon is approached from different angles. It can be safely said that none of these theories is necessarily better than any other, nor explains post-Independence nationalism in a fool-proof manner. Also, it should be noted here that, the commonly accepted concept about a nation-state is no longer valid. The classical concept about a nation-state is that of an ethnic group having their own independent politico-economic system. But this type of ethnic homogeneity was and still is very rare. That is why Anthony Smith opines that the very term 'nation-state' is a misnomer. [14]

Ethnic pluralism rather than ethnic homogeneity appears still to be the norm. Most of the states in the world today are composed of two or more ethnic communities. Both large and small states possess sizeable minorities and most states have small ethnic minorities. Centuries of conquest and population movement resulted in the geographical intermingling of many cultural groups in a complex pattern. So the nationalist image of the world consisting of states based on culturally distinctive groups occupying homogeneous, compact and mutually exclusive national territories is far from reality. [15] Hugh Tinker and Cornelia Navari also subscribe to this view. [16] Breuilly also observes that measuring cultural homogeneity or heterogeneity is very difficult. [17]

These facts and opinions point towards the necessity of

defining nationalism functionally. The post-independence phase of nationalism can be understood fully, only if one defines it, according to the 'statist' tradition. Hence, nationalism may be defined as acceptance of the state as the impersonal and ultimate arbiter of human affairs by a people who live together in a common territory, cherish a common heritage, and wish to live together now and in the future in their own state. [18]

This definition gives due regard to the fact that nationalism is supported by and expressed through power factors. Any movement or activity that does not accept the state for the settlement of disputes cannot be nationalist. Such movements can be termed as 'communalist' or 'separatist'. Geertz also agrees with this view when he states that anti-state movements are merely communalist or separatist secessions, based on the 'primordial sentiments' of race, language, religion, etc. [19] In his opinion, nationalist movements are essentially state-oriented and state grounded ones. [20]

For a clear understanding of nationalism, the term sub-nationalism also has to be defined. In this study, it denotes the utilization of group loyalty to primordial sentiments as a resource for the attainment of political ends.

Nationalism involves a two way process. The people accept the state as the final arbiter of their conflicts. Thus, they express their loyalty to the state. The state resolves human conflicts through the institutional mechanism of its power in such a manner that it leads to consensus generation. Conflict resolution and consensus generation are the main functions of nationalism in an independent society.

It has to be noted that during the period of anti-colonial nationalist movement, there were innumerable local and regional level conflicts, particularly related to identity. It was humanly impossible to solve these and create a national identity in place of those. The heterogeneous nature of the society was the main reason. So it was natural that those primordial loyalties continued even after independence.

Conflict is inherent in such a society of diversity. But this does not mean that cultural homogeneity would prevent conflicts. Breuilly observes that based on the circumstances, even culturally homogeneous people can divide on the basis of trifling cultural distinctions. Just like that, a people divided by fundamental cultural distinctions can cooperate politically within the same state. [21]

Moreover, most often, the conflicts in a plural society are a part of the politics of bargaining. So, if the political system has enough wisdom and calibre to take bold decisions and yet be accommodative to reasonable demands, it can not only positively restrain the conflicts that aim at unfair and profligate bargaining but also convert them into hale and hearty partnership deals in the interest of nation-building. [22] The system should be firm because a nation which attains independence must protect its sovereignty. This does not mean that the system should try to assimilate the minorities to the majority. The minorities need not give up their ethnic, cultural and linguistic characteristics. The growth and development of nationalism in an independent state does not necessarily require an erosion of primordial loyalties. [23]

Even while having loyalty to primordial sentiments, people can have shared positive affections toward the nation-state. Such positive affections originate out of the feeling of the people that their nation-state is just and stands for the equal benefit of all of them. This feeling is exhibited by the people through their acceptance of the state as the final arbiter of their disputes.

Moreover, when the involved groups feel that their voice and their aspirations are given due weight and that they have a say in the political system of the nation, it helps the system to get the consensus of the groups. This two-way process of resolving conflict and receiving consensus strengthens nationalism. The main political functions of nationalism in an independent, culturally plural society are, as stated earlier, conflict resolution and consensus generation. Conflicts can be resolved or minimized by adopting two methods: assimilation and accommodation. Assimilation is the process by which different groups drop their cultural and ethnic distinctness and identities, and are immersed in the dominant group or a new group formed from the component parts. [24]

Since groups have a tendency to preserve their distinctness, official policies aimed at assimilation may lead to conflict situations. Therefore, many plural societies prefer accommodation to assimilation. Accommodation is the process through which people belonging to different subcultures and ethnic groups cooperate for mutual benefit in spite of differences or hidden hostility. For achieving accommodation among different cultures and ethnic groups, the sources of inter-group conflicts should be reciprocally attuned. This would enable each group to enjoy reasonable freedom and security without surrendering all differences. Consensus can be generated or increased by adopting two methods: one by adoption of positive nation-building techniques and two, conflict resolution through accommodation.

Based on these theoretical premises, one should look at the major conflicts faced by the Indian political system during the post-Independence phase, analyse how these were resolved and assess the impact of these on the wider political spectrum. After Independence India had to confront sub-national movements in few places like Tamil Nadu, Nagaland and Mizoram. But the strength of Indian nationalism helped to solve these amicably and bring the people of these places to the national mainstream. In the present work, three important problems related to sub-national loyalties are studied. These are three major conflicts related to the people's loyalty to language, religion and region that have challenged the strength of Indian nationalism after Independence.

To start with, we may look at the pan-Indian movement for reorganization of states on the basis of language. A brief explanation of the causes of this movement would not be out of place. Firstly, there is the fact of linguistic diversity as revealed by the Census reports which categorise 1,652 spoken languages and dialects in India. Several migrations like that of the Dravidian and Aryan races and the advent of the Turks, Iranians and Mughals as well as the English have contributed to linguistic diversity. Secondly, the agitation over the partition of Bengal starting in 1905 added a political dimension to the language problem. Thirdly, Constitutional changes as envisaged in 1911, and in the Government of India Acts of 1919 and 1935 created new linguistic units. Fourthly, the Congress session of 1920 divided India into 25 linguistic units for purposes of party organization. Fifthly, the Congress resolved in 1927 to recommend creation of some linguistic provinces. Sixthly, there was a reaction against creation of linguistic states, as evident from the Dar Commission Report of 1948, which led to more agitation. Seventhly, there was the fast and death of Potti Sri Ramalu of Andhra in connection with the creation of the Andhra State which compelled the Government to yield to the demand for the creation of Andhra State. Finally, there was the States Reorganization Commission Report of 1955 followed by legislation to reorganize states. The process started in 1956 has been continued by the creation of Tribal states in the North-East, partition of Punjab and by granting full-fledged statehood to several Union Territories.

Many people including Jawaharlal Nehru were afraid that demands based on the language loyalty of the people of India were capable enough to challenge the unity of India. But time proved that their fears were baseless. It took some time for the political leadership of the nation to learn that this demand was a reasonable one and not secessionist. The reorganisation of states mainly on linguistic lines led to successful conflict resolution. Here, the people's acceptance of the State as the ultimate arbiter of their affairs was clear. So this demand cannot be termed as separatist. The States reorganisation had, in fact, enabled all sections of the population to participate in the Government of the country by overcoming linguistic barriers.

Thus, the reorganisation promoted the unity of India.

Another crisis that the country faced after Independence was related to sub-national demands based on regionalism. The intricacies of the problem of sub-nationalism in India is more reflected by the North-Eastern states—Arunachal Pradesh, Assam, Manipur, Meghalaya, Mizoram, Nagaland and Tripura —than by any other part of the country. Most of the States are very small in size and population. The total population of the region is only 29 million. However, the ethnic, linguistic, religious and cultural diversity of this population is glaring. There are 116 scheduled tribes in the region. They belong to at least five different ethnic groups. Religious diversity is reflected by the presence of Hinduism, Buddhism, Islam, Christianity and certain traditional religious practices. Linguistically, the region has Assamese, Bengali, Manipuri and a variety of local languages. English is the official language in Nagaland, Mizoram and Meghalaya.

Economically, North-East is better than many Hindi speaking states like Uttar Pradesh. However, the strategic location of the region and the nature of the polity, economy and society of the neighbouring regions and countries have added a set of special concerns, problems and anxieties to its normal national share. Thus, the politics of this region has been deeply influenced by the influx of people from other regions as well as neighbouring countries like Myanmar and Bangladesh. A typical example is provided by the 'citizenship crisis' of Assam.

This agitation, started under the leadership of the All Assam Students Union (ASSU) and the All Assam Gana Sangram Parishad (AAGSP) in 1979 with the main demands of detection, disenfranchisement (or deletion) and deportation of foreigners who migrated to Assam, was resolved in August 1985. Migration has been an important contributory factor to Assam's abnormally high rate of population growth. The economic, social and political superiority achieved by the migrants over the natives was a factor that provoked the agitators. The relative economic backwardness of the State, the fear of the Assamese people of losing their identity etc., caused the agitation.

The issue remained unresolved for almost six years. In the initial stages of the agitation, the leaders of the agitation as well as the Union Government took an adamant attitude and were not prepared for reconciliation. The youthful, idealistic, and ambitious leaders of the agitation were not ready to move even an iota from their stand. At the same time, the intelligent, experienced and diplomatic representatives of the Union Government were expected to win over the "boys" easily. This created a stalemate which helped the extremists to apply terrorist tactics to win the cause. The State and the Union Government were not to be cowed down by such tactics. Talks to resolve the issue very often failed and the agitation continued for a long time. A number of factors such as time, the weakening of the forces of agitation, the agitators losing their hold over the people, the agitation leaders' realisation that certain demands such as the demand for the deportation of all foreigners were unattainable or that even if the Government conceded that, the actual deportation would take not only years to complete but was also virtually impossible.

The Government's policy of attempting rapid economic development of the State, the desire of the new leadership of the nation to solve the issue immediately, etc., paved the way to the accord. An important aspect that is to be considered here is the nature of the agitation whether it was anti-Indian or not. There was criticism that it was "an open movement for secession". [25] But there is lack of data to substantiate it. The agitators did not seek or receive help of any foreign nation. The agitators have no proven extra-territorial loyalty. They never demanded sovereignty or separate statehood. Thus, it is clear that the agitation was not aimed at creating any anti-Indian sentiment and hence, was not anti-national. Moreover, once the problem was solved, almost all the leaders and their followers expressed their willingness to come to the mainstream of Indian nationalism.

This predicament was not one precipitated by fissiparous tendencies. The people of Assam turned to the Central

Government for correcting the wrongs done to them by various governments that came into power in Delhi since the time of the Colonial rule. The accommodative tactics applied by the Central Government for conflict resolution was successful. Though the regional party that projected the Assamese sub-nationalism came into power in the State following the signing of the Assam accord, the people rejected it in the next general elections and voted for a Congress (I) Government. This was a vote against confusion, instability and violence.

The strength of Indian nationalism was challenged also by what was called 'the Punjab problem' which is the term used to connote the demand for Khalistan and connected activities, for a few years. The Akali Dal, the most important regional political party of Punjab, has a prominent place in the politics of the region. The political ambitions of the Akali politicians enabled or rather forced them to adopt and apply almost all available strategies including terrorism to attain their goals. The politics of the Sikhs starts with the successful attempt of the Akali Dal to control Sikh temples and Gurudwaras as early as 1920. The Gurudwara Reform Movement paved the way to the birth of the Shiromani Akali Dal on December 14, 1920.

Initially, the Akali Dal identified itself with the Indian struggle for freedom. In the second stage, the Akali Dal becomes a provincial political party and is given a place in the non-Congress Unionist ministry of Punjab. In the third stage, the Akali Dal opposes the creation of Pakistan and even demands a separate nation for the Sikhs in the event of partition. The fourth stage commences with the Akali agitation for a separate state—the Punjabi Suba—with a Sikh majority. The fifth stage marks the success of the agitation and the creation of the Punjabi Suba with 55 percent Sikhs. The sixth stage marks the clash between the Akalis and the Congress for the control of the new state and the formation of the first Akali Dal ministry. This is followed by the elections of 1977 when the Akali Dal joined the Janata Party in setting up a non-Congress government in Punjab. The eighth stage marks the advent of the leadership of Bhindranwale and the demand for an independent Sikh state of 'Khalistan' followed by the military action of the Centre. The ninth stage marks the accord of Rajiv Gandhi with the Longowal faction of the Akali Dal and the formation of the Barnala ministry. The tenth stage is again of Central rule in Punjab marked by terrorism and the increase in Sikh militancy. The eleventh stage marks the 'return of normalcy' and elected Governments.

Political, economic, social, religious and linguistic problems are intertwined to make the Punjab problem one of the most complicated and hard-to-solve problems faced by the Government of India since Independence. However, it cannot be characterized as a problem created out of the divisive mentality of the people of Punjab. An analysis of the various aspects of the problem shows that, though sometimes a minority demanded the establishment of a separate nation, the majority was always against it. The latter's demands included more autonomy, more economic opportunity and more respect to the Punjabi religion and language.

The people of Punjab compared to the other states in India, form an affluent society. Their expectations for more agricultural and industrial development and more employment opportunity for their industrious and hard working youth could not be realized to their satisfaction. Neither the State Government nor the political parties of the State were ready to take the responsibility. They, instead, vied with one another in blaming each other. The Central Government also was blamed for putting hurdles in the path of the State's economic development. The Akali Dal, while in power in the State, blamed the Congress ministry at the Centre and the Congress party of the State for the problems of the State. They did the same while they were out of power.

However, while they formed coalition ministry in the State with the Congress, they were busy with preserving their power position rather than criticizing the Congress or the Centre. However, all aspirants of power of a party could not be accommodated in the ministry. A few of those who were out of power formed their caucus and criticized the coalition ministry as well as the Centre. Such groups could exploit the political, economic, religious and linguistic aspirations and sentiments of the people. It could be noted that this was often resorted to achieve the ambitions of politicians and that this was based on no great principles.

The religious, ethnic and linguistic criteria are used as political resources by the Sikhs to substantiate a claim to sub-national identity. It was in order to effect a change in the distribution of resources, including political power that the Sikhs appealed to sub-national loyalty. But a minority, with the connivance of certain foreign countries made use of this opportunity to undermine the integrity of India. Many factors such as the geographical setting of the region, the growth of religious fundamentalism, etc., made the State vulnerable to such foreign influences. But the terrorists and their secessionist demands were not supported by the majority of the population of Punjab. In fact, the majority has shunned terrorism and expressed the desire to be in the national mainstream.

These studies point towards certain facts. Firstly, in post-Independent India, nationalism has become a stronger force than sub-nationalism which is the creation of primordial sentiments. Secondly, it was the people's desire to live together and maintain their state that was expressed through their acceptance of the state as the final arbiter in the case of states reorganisation, the Assam agitation and the Punjab problem. Thirdly, in all these cases the political system helped to resolve the conflicts by adopting the method of accommodation. On many occasions, the people had expressed their approval of the state's approach to conflict resolution. Surely, this is the reason why the people turn more and more to the state for conflict resolution. Lastly, in India, people can have a belief in linguistic, religious or regional sub-nationalism and an Indian nationalism at the same time. There is currency for the term 'Hindu nationalism' particularly after the coming into power of the Bharatiya Janata Party (BJP).

Unifying Factors

There are many factors that prevent sub-nationalism from becoming secessionism in India. The most important factor is the fundamental unity of India about which volumes have been written. Another one is contributed by the deliberate and planned measures pursued by the political system of the nation with the help of a written Constitution and a functioning democratic political party system. The political system is fashioned in such a way that it can accommodate variety and multiplicity. People's involvement as well as faith in the political system is another factor that strengthens nationalism. This is explained by the tremendous popular-participation in the democratic process. Yet another factor is the political economy of the nation. Rapid economic development especially since 1991 has created a large middle class with a stake in Indian unity. People are now hopeful that "The inversion between capitalism and democracy suggest that India might have a more stable, peaceful, and negotiated transition into the future than, say, China". [26]

The nature of the political leadership is another factor that deserves mentioning. While in many states the leadership is unstable, at the national level there has been the tradition of strong leadership. The plural nature of the Indian society has a very important role in preventing sub-nationalism from becoming secessionism. While at the national level, India is a federal society in which language groups and religious minorities have an important role, each state is a plural society in which castes and communal groups play a political role. [27] Many unilingual states are plural in many other ways. Actually, not even a single state is completely unilingual. Linguistic and religious overlapping is quite common. These are some of the factors that make sub-nationalism a weaker political force than Indian nationalism.

Apart from these, there are many other integrating forces in

India such as the educational system, the introduction of modern technology and the important role of the communication media. In spite of the language barriers, the educational system tends to unify the country by subjecting the students to a process of socialisation. The media such as television and radio are powerful instruments of integration. Though regional newspapers may sometimes support secessionist demands, the press, taken as a whole, has been a powerful agent of unification by compelling the country to think in terms of the nation.

Conclusion

India today is a unified, strong and vibrant nation. It is evoking an 'intense belief in its future in the minds of many like Nandan Nilekani, who writes, "I still think that after a long and convoluted path, after many a stumble and wrong turn, a different kind of moment seems to be upon us. For the first time, there is a sense of hope across the country, which I believe is universal. There is a momentum for change, evident in the enthusiasm of our younger legislators, the mushrooming of civil society organizations, and activists fighting in the courts for reforms in governance and for the protection of fundamental rights". [28] These fighting in the courts even for reforms in governance are testimony to the fact that people consider their state just and hence the final arbiter of their affairs. This in turn is thanks to the political process of accommodation of diversity that the political system is pursuing and which makes nationalism a living force in India. Nationalism will continue to be a strong, vibrant and positive force in India as long as the political system resolves conflicts in a manner that generates consensus.

Endnotes

1. Arthur. N. Waldron (1985), "Theories of Nationalism and Historical Explanation", *World Politics*, 37, April, p. 416.
2. Carlton J.H. Hayes (1928), "Essays on Nationalism", New York, p. 94.
3. T.V. Sathyamurthy (1983), "Nationalism in the Contemporary

World: Political and Sociological Perspectives", London, pp. 6-16.
4. Ernest Gellner (1978), "Thought and Change", Chicago, p. 56.
5. Anthony D. Smith (1971), "Theories of Nationalism", London, pp. 145-148. For a criticism or Gellner's approach see, Authur Wlaldron, op. cit. (n.1), p. 422.
6. Karl W. Deutsch (1966), "Nationalism and Social Communication", Massachusetts, p. 97.
7. John H. Kautsky (1962), "Political Change in Underdeveloped Countries: Nationalism and Communism", New York, pp. 13-56 and Peter Worsley (1970), "The Third World", Chicago.
8. For details see, S.N. Elsernstadt (1966), "Modernization: Protest and Change", Englewood Clifts; N.J. Smelser (1968), "Essays in Sociological Explanation", Englewoods Cliffs and K.H. Silvert (1983), "Expectant People: Nationalism and Development", New York.
9. Albert Breton (1964), "The Economics of Nationalism", *Journal of Political Economy*, Vol. 72, August, pp. 376-386; Harry Johnson (1965), "A Theoretical Model of Economic Nationalism in New and Developing States", *Political Sciences Quarterly*, Vol. 80, June, pp. 169-186.
10. For details see, John H. Kautsky (1962), "Political Change in Undeveloped Countries: Nationalism and Communism", New York; T.B. Botto-more (1965), "Classes in Modern Society", London.
11. Gordon Lightbody (1964), "A Note on the Theory of Nationalism as a Function of Ethnic Demands", *Canadian Journal of Political Science*, II, Summer, p. 331.
12. Georges Balandier (1970), "The Sociology of Black Africa", London.
13. For details see, M. Gluckman (1966), "Custom and Conflict in Africa", Oxford, and M. Gluckman and F.R. Jevons (eds.) (1966), 'Closed Systems and Open Minds', London.
14. Anthony D Smith (1981), "The Ethnic Revival", Cambridge, p. 10.
15. A.W. Orridge (1981), "Variates of Nationalism", in Leonard Tivey (ed.), 'The Nation State: The Formation of Modern Politics', Oxford, p. 41.
16. Hugh Tinker, "The Nation State in Asia", in Leonard Tivey (ed.), Ibid., p. 119.

17. John Breuilly (1982), "Nationalism and the State", Manchester, p. 223.
18. This definition, in a way, combines the statist definition of nationalism given by K.H. Silvert, and the concept of nation as defined by Boyd Shafer. For details see, K.H. Silvert (1983), "Expectant People: Nationalism and Development", New York and Boyd Shafer (1972), "Faces of Nationalism", New York.
19. Clifford Geertz (1963), "The Integrative Revolution: Primordial Sentiments and Civil Politics in the New States", in Clifford Geertz (ed.), 'Old Societies and New States: The Quest for Modernity in Asia and Africa', New York, pp. 110-111.
20. For a study related to nationalism defined on the basis of territoriality, see, Jaun J. Linz (1985), "From Primordialism to Nationalism", in Edward A. Tiryakian and Ronald Rogowski (eds.), 'New Nationalism of the Developed West: Towards Explanation', Winchester.
21. John Breuilly (1982), "Nationalism and the State", Manchester, p. 223.
22. Iqbal Narain (1976), "Cultural Pluralism, National Integration and Democracy in India", *Asian Survey,* Vol. XVI, No. 10, October, p. 903.
23. Emerson holds the view that nationalism can foster a sense of common purpose only if it demonstrates the ability to overcome primordial sentiments. Refer, Rupert Emerson (1969), "The Problem of Identity, Statehood and Image in the New Nations: The Situations in Africa", *Comparative Politics*, April 1, p. 306.
24. Mburbumba Kerina (1978), "Plural Societies and the Application of Democracy", in Nic Rhoodie (ed.), 'Inter-group Accommodation in Plural Societies', London, p. 307.
25. For details see, P. Ramamurthi (1980), "Real Face of the Assam Agitation", New Delhi.
26. Gurcharan Das (2007), "India Unbound: From Independence to Global Information Age, Penguin Books, New Delhi, p. xix.
27. V.K. Sukumsran Nayar (1966), "New Challenges before the Indian Federalism", *Kerala Journal of Political Studies*, Vol. 1, No. 1, January, p. 93.
28. Nandan Nilekani (2008), "Imaging India: Ideas for the New Century", Penguin, New Delhi, p. 484.

5

Human Security, Human Development and Human Rights: A Perspective of India's North-East

Haans J. Freddy

Introduction

Two dominant strands of foreign policy during the 1990s, viz. economic development and military security became intertwined. With the end of the Cold War, both development and security concerns which underwent conceptual turmoil were complemented with highly uneven patterns of change in different pieces of development along with technological advancements and political changes which are often labelled as globalisation. Thus, emerged the concept of human security as a consequence of such developments (King and Murray, 2002:585).

Loosely defined as people's freedom from fear and freedom from want in a broad sense, the human security approach presents an opportunity to investigate many issues relating to human development and human security. In 1994, the global Human Development Report introduced the idea of human security around which a wide array of literature and initiatives have been built upon. Broadening the scope of security analysis and policy from territorial security to the security of people, the United Nations General Assembly (UNGA) Resolution in the year 2012, stresses the importance of member states to identify and address widespread and cross-cutting challenges to survival, livelihood and dignity of their citizens (Gomez and Gasper, 2013; Gasper, 2005:221).

Human security being a critical component of the global political and development agenda, firstly, is directed towards the protection of the individual, encompassing a strategic concern for both national and international security and secondly, it shifts its focus from traditional security issues relating to matters of national defence, law and order to political economic and social issues enabling a life free from fear and risk (Hussein, Gnisci and Wanjiru, 2004:8). While the concepts of security and insecurity have relative connotations, insecurity for some could mean loss of guarantee of access to jobs, health care, social welfare, and education; for others it could mean violation of human rights, extremism, domestic violence, spread of conflicts, displacement, etc (Tadjbakhsh, 2009).

Attracting interest, approval and critique from the academic community, human security proponents see it as a timely and essential expansion of traditional security concerns and as a useful tool for shaping foreign policy (Oberleitner, 2002). Despite claims which assert that human security represents a new paradigm, it remains unclear whether the concept can contribute to academic research or governmental policy making. Human security has two problems within its discourse. First, human security seems like sustainable development and everyone is for it but the concept lacks a clear definition and is too expansive and vague, covering everything from physical security to psychological well-being, which provides policymakers with little guidance in the prioritization of competing policy goals and academics little sense of what, exactly, is to be studied. Second, the idea of human security glues together a host of issues which seek to shift attention and resources away from conventional security issues and toward goals that have traditionally fallen under the rubric of international development. However, the concept is slippery as it diminishes the concept's usefulness as a guide for academic research or policymaking (Paris, 2001:88).

Definitions of Human Security

Although there has been repeated indication that the idea of human security lacks a clear definition, proponents of a broad definition have suggested that there should be a focus on what the act of defining within security studies inherently articulates about power relations based upon political, moral and ethical choices. The lack of an agreed definition of human security does not mean conceptual weakness but rather showcases the refusal to accept the difficulties of the discipline's definitional project (Grayson, 2004:357). Thus, a broad definition of human security is therefore critical to raise questions that are significant to security studies that relate and apply to the everyday lives of humans (Tadjbakhsh and Chenoy, 2007:39).

The simplest definition of human security could refer to the absence of insecurity and threats. The capacity to deal with and identify threats and to avoid them wherever possible and to mitigate their effects encompasses the concept which goes beyond the traditional idea of physical security to the guaranteeing of people's livelihoods against sudden calamities (Ibid.:40).

"Human security refers to the quality of life of the people of a society or polity. Anything which degrades their quality of life—demographic pressures, diminished access to or stock or resources, and so on—is a security threat. Conversely, anything which can upgrade their quality of life—economic growth, improved access to resources, social and political empowerment, and so on—is an enhancement of human security" (Thakur, 1997:53-54). "In policy terms, human security is an integrated, sustainable, comprehensive security from fear, conflict, ignorance, poverty, social and cultural deprivation, and hunger, resting upon positive and negative freedoms" (Ginkel and Newman, 2000:79). "The term human security focuses on the concept of security on human survival, wellbeing and freedom". "...we conceptualize human security as the objective—the ultimate ends—of all security concerns. In this schema, other forms of security, such as military security, are not ultimate goals. Rather, these other forms of security are simply means for achieving the ultimate objectives of human security" (Chen, 1995:139).

According to both 'critical' and 'human' security

approaches, security is about attaining the social, political, environmental and economic conditions conducive to a life in freedom and dignity for the individual (Hammerstad, 2000: 395). Human security can be said to have two main aspects. It means, first, safety from such chronic threats as hunger, disease and repression. And second, it means protection from sudden and hurtful disruptions in the patterns of daily life—whether in homes, in jobs or in communities. Such threats can exist at all levels of national income and development. The list of threats to human security is long, but most can be considered under several main categories: (i) economic security, (ii) food security, (iii) health security, (iv) environmental security, (v) personal security, (vi) community security and (vii) political security (UNDP, 1994:23-25).

For Canada, human security means freedom from pervasive threats to people's rights, safety or lives. Canada has identified five foreign policy priorities for advancing human security: (i) protection of civilians, concerned with building international will and strengthening norms and capacity to reduce the human costs of armed conflict; (ii) peace support operations, concerned with building UN capacities and addressing the demanding and increasingly complex requirements for deployment of skilled personnel, including Canadians, to these missions; (iii) conflict prevention, with strengthening the capacity of the international community to prevent or resolve conflict, and building local indigenous capacity to manage conflict without violence; (iv) governance and accountability, concerned with fostering improved accountability of public and private sector institutions in terms of established norms of democracy and human rights; (v) public safety, concerned with building international expertise, capacities and instruments to counter the growing threats posed by the rise of transnational organized crime. [1]

Japan emphasizes "human security" from the perspective of strengthening efforts to cope with threats to human lives, livelihoods and dignity such as poverty, environmental degradation, illicit drugs, transnational organized crime, infectious diseases such as HIV/AIDS, the outflow of refugees and anti-personnel land mines, and has taken various initiatives in this context. To ensure "human freedom and potential", a range of issues needs to be addressed from the perspective of human security focused on the individual, requiring cooperation among the various actors in the international community, including governments, international organizations and the civil society. [2]

Thus, it clearly implies the need to assess human security in terms of its contribution to changing normative contexts and the changing perceptions of actors' identity and interests. This means that it is important to shift the focus from looking at human security as a policy agenda and assess it as a security discourse which both reflects and works towards changing the important contexts where different conceptions and practices of security are evoked by actors. It is then that an assessment of human security would be consistent with the way security actually operates in international politics.

Human Security and Human Rights

Human rights which has occupied a central place in various interesting debates and discussions both at the domestic and international levels is connected to the idea of human security. This section looks to examine to what extent the idea of human security adds to the concept of human rights and the discussion surrounding it. Although there seems to be a close relationship between human security and human rights, this aspect is yet to be analysed comprehensively. This is evident from the fact that both human rights and human security have a focus on the human individual. Oberleitner (2002:14-15) raises three important questions in this regard: (i) is human security a human right itself?; (ii) are human rights the core and normative base for human security? and (iii) what is the relation between human security and human rights in terms of mutual enrichment as well as potential mutual dilution of the respective concepts?

While human rights are always concerned with human security, it is to a certain extent clear that security itself is a right (ibid). Article 3 of the Universal Declaration of Human Rights (UDHR) has three inter-linked rights: (i) the right to life, (ii) the right to personal freedom and (iii) and the right to personal security. This is often seen in relation to Article 5 of the UDHR which provides for the prevention of torture and Article 9 which speaks about the prevention of arbitrary arrest or detention. Such provisions for human security can be seen in various instruments such as the UDHR, International Covenant on Civil and Political Rights, International Covenant on Economic Social and Cultural Rights, The American Declaration on the Rights and Duties of Man, The African Charter on Peoples and Human Rights, and The American Convention on Human Rights. An attempt to locate security in human rights instruments is an indicator which answers the question whether human security is indeed a human right.

Human rights and human security do not possess the same meaning nor do they overlap. Although they have different meanings and functions, there is a possibility that there are strong arguments made towards linking the two concepts. Governments' increasing involvement in the human security network and also the interest shown by the Independent Commission on Human Security (CHS) are in fact promising efforts towards this linkage. The report 'Human Security Now' prepared by the CHS has suggested that human security and human rights are complementary to each other (Boyle and Simonsen, 2004:5-6).

The idea of human security fits well with the concepts of human rights and human development. Human security is concerned with reducing and removing the insecurities that plague human lives. It is in contrast with the notion of state security which concentrates on the integrity of the state while only having an indirect connection with individuals living within the states. The contrast may be clear but while delineating the idea of human security, it is important to identify how human cantered concepts such as rights and development relates to and differs from human security. These concepts are widely known and have been for long time championed as those having a direct concern with human lives. Human security fruitfully supplements the expansionist perspective of human development by directly paying attention to the downside risks. The idea of human security demands that humans are protected from the insecurities that threaten the safety or survival, the natural dignity of men and women, human beings exposed to diseases and sudden dangers that may afflict them. It also seeks to provide empowerment of people so that they can cope with and overcome these difficulties.

In this context, it is clear that there is of course no contradiction between the concepts of human security and human development. However, the emphasis and priorities that are placed in the human security perspective are quite different from those that are found in the upward-oriented literature of the human development focus which tends to emphasize on growth and equity. In contrast, human security focus tends to concentrate on the downturns with security, since downturns occur when there is global or local affliction. This occurs when there is persistence of insecurity caused even when there is growth coupled with uneven distribution of benefits. The economic case only presents the contrast between the expansion with equity and the downturns with security.

In similar terms, the two concepts human security and human rights complement each other. The idea of human rights which has a universal appeal has been quite attractive in that every individual or citizen in any part of the world has certain basic rights that others should respect. This moral obligation has been used in different contexts from resisting torture and arbitrary incarceration to demanding the end of hunger, poverty and the unequal treatment of women. The idea of human rights which rests on the premise of demands/claims (which are very strongly supported) looks to aid and enhance certain basic rights or freedoms. This basic normative nature of the idea of human rights leaves us the question as to which rights and freedoms are crucial that requires the state/society to acknowledge, safeguard and promote. It is in this context that the idea of human security can make a crucial contribution by identifying the importance of freedoms from basic insecurities.

Such considerations can make security crucial and can join hands with the ethical force of claims for the recognition of certain freedoms which human rights guarantees. Thus, while on the one hand human rights is seen as universal and general in nature which needs to be filled with specific demands/claims, it is significant that human security fills this gap through reasoned substantiation. It (human security) is also important since the demands made through claims (human rights) need political recognition, it is important to note that this can be appropriately obtained by seeing certain freedoms as an important class of human rights. This can be argued in the context of the associative connections which human security and human rights have with the corresponding duties other people and institutions have. Duties can take the form of perfect and imperfect obligations. Perfect obligations constitute specific demands on particular agents or persons, while on the other hand, imperfect obligations constitute general demands on anyone in a position to help reduce insecurities in human lives.

Thus, looking at human security within a general framework of human rights, can be highly productive. Human rights and human security are thus mutually reinforcing. Human security identifies rights that are in stake in a particular situation and human rights helps answer the question how human security in that particular situation can be promoted. The notion of obligations and duties complements the recognition of the ethical and political significance of human security (Commission on Human Security, 2003:9-10).

Conflict in North-East India

India's North-East encompasses a tremendous social heterogeneity and complex traditional political organizations. With the existence of numerous cultural groups with separate histories, the region is home to 75 ethnic groups and approximately 400 languages or rather dialects (Dev, 2004:4749). North-East India is most commonly studied by scholars as a theatre of insurgency and counter insurgency. A high degree of mutual alienation marks the relationship between the population of North-East India and Indians on the other side of the narrow corridor that connects the region to the rest of the country. Integration without consent, colonial attitudes, nativism, legal and illegal migration relative deprivation, cultural nationalism and irredentism have sparked violent conflict in the region for over five decades.

Numerous peace accords have been signed between the state governments and the Government of India and insurgent groups to bring an end to the conflict. An examination of the thirteen peace accords signed to bring lasting peace in the region have barely shown any characteristics of moving towards this direction. Pre-accord talks have not been inclusive, the provisions agreed upon with one group frequently and patently conflict with the interests of others and the accords either contain provisions to address core conflict issues in a manner that cannot be implemented or they do not deal with the core issues at all.

Outstanding conflicts and the continuing ability of the United Liberation Front of Assam (ULFA) to operate and raise questions about political will and sincerity envelops all other dimensions as well. In North-East India, numerous peace accords have provided for the redistribution of authority by either creating territorial or non-territorial means of representation and self-governance—a time honoured conflict resolution measure. One useful lesson from the patterns of peace accords in North-East India is that when many administrative levels exist through which a group's status can be upgraded, the parties to the conflict have more room for negotiation. On the other hand, devolution of power remains a real problem in the region, and creating new administrative or territorial units is not a guarantor of viability, resource independence or autonomy (Rajagopalan, 2008:ix-xi).

The British administered India's North-East as an imperial frontier and treated it as a buffer zone between Bengal plains and the highlands of China and Burma. They avoided imposing direct administration over much of the multi-ethnic, polyglot area with only Assam witnessing the maximum British investments in tea, oil and timber. After the British left, India sought to integrate the region into the national mainstream, provoking violent opposition to the federal control and assimilation. Guerrilla warfare emerged as the favourite form of resistance and seventy years after Independence of India, more than seventy insurgent groups still remain active in North-East India. They continue to battle the Indian security forces for independence although some have settled for extensive autonomy. The ethnic rebel armies often fight each other conflicting mainly over homeland demands and scarce resources (Bhaumik, 2007:4).

Ethnic conflict began in North-East with the Assam-Nagaland border dispute. When Nagaland was formed in 1963, tension between Assam and Nagaland over the boundaries for over four decades has prevailed between security forces as well as between ordinary civilians. Every state in the region except Sikkim is currently affected by some form of insurgent violence and four of these (Assam, Nagaland, Manipur and Tripura) have witnessed scales of conflict that could at least between 1990 and 2000—be categorized as low intensity conflicts in which fatalities have been ever increasing (Das, 2007:1).

Conflicts in North-East India have received much attention from scholars and it has been analyzed in the context of ethnic insurgencies and the response of the state in terms of containing (counter-insurgency) these movements. While the insurgent groups in the region have asserted their demands through violent methods/guerrilla warfare, they also receive considerable support from the local population. The support received from the local population is also due to the fact that many of the individuals who join these groups are members of the family of a particular group. In other words, members of the insurgent group are ethnically attached to the ethnic group which they represent which is also in the form of family attachments. It is also interesting to note that much of the members of the ethnic groups are youth who have joined these groups either because they have been passive or direct victims of human rights violations in the region. [3]

It is in this context that in recent times, the concept of human security when applied to the North-East of India poses questions which are not only interesting but also important.

Human Security in North-East India

Intellectual responses to terrorism in India has been marked by a broad stream of justifications and this tendency persists even after decades of strife which has cost many innocent lives in the nation and billions of rupees towards counter insurgency initiatives and billions more with respect to developmental activities. To this, there has been a proliferation of a false sociology of poverty, political neglect and whatever other stereotypical pop-psychological explanations. At its worst, is a propaganda which is a source generated by the terrorist organizations in the region and on the other, on an unexamined dogma that reduces all conflict and its resolution to developmental catechisms. This intellectual fashion has had a powerful, if fitful, influence, both on India's national policy, and on the pattern of demands raised by the governments (and indeed by the terrorists themselves) of various states afflicted by the scourge of terror.

The result has been a massive—though usually *ad hoc* and erratic—allocation of resources by the Centre to such states, particularly at times when a central government is under some political pressure to be seen to be responding 'effectively' or 'proactively' to terrorism. In all such cases, little effort has been devoted to evaluating the utilization or impact of such allocations in the past, and these actions are simply premised on the erroneous assumption that ending grievances will end terrorism. The dynamics of terrorism, however, acquire a character and autonomy entirely divorced from the paradigms of such naive determinism, resulting in a high rate of failure for all initiatives based on the conviction that terrorism can be made to disappear through appropriate developmental investments. Indeed, the only things that have been known to 'disappear' in most such cases, are all signs of the funds invested, and it is somewhat surprising that this dogma has persisted for such a long time in the complete absence of supporting evidence (Gill, 1998).

The concept of security dilemma in international relations is used to explain the anarchic nature of global politics. In the absence of an over-arching authority, according to this theoretical tradition, sovereign states are forced to provide for their own security through self-help, in turn causing the insecurity of other states. In situations where one ethnic group is organized as a national liberation force threatening the survival of another rival group, the state is often seen as a reliable provider of security and it is easy to see why the group which is facing threat would look to self-help as a means to providing security by forming its own armed militant group. Mainland India may not be capable of thinking that the Indian state is incapable of providing security to citizens in many parts of the North-East. It should also be noted that no state can provide security everyday through military might alone.

Moreover, what seems as extremism, insurgency or militancy may not be so in the local context as it may seem in the nation's capital. In other words, a group which might seem as an extremist might be a reliable source of security from the perspective of its ethnic constituency. Indeed, in an ethnically divided situation, where the actions of Indian security forces may be seen as partisan, offensives against militants who are seen as security providers by their ethnic kin, may even add to the latter's sense of insecurity and an incentive for strengthening the self-help form of security. The perceived effectiveness of militant groups in providing security to their ethnic kin—at least compared to that of the state—is quite self-evident to their followers and supporters. If we assume that citizens are smart enough to vote with their feet, the very proliferation and persistence of small bands of militant groups in the face of the long and violent history of counter-insurgency operations would suggest that militant groups persist and proliferate only because they serve important functions. Their incapacity to deliver on the agenda of 'national liberation' may be quite irrelevant to the politics of security in a context where the state cannot guarantee it (Baruah, 2002: 4179-4180).

People of the North-East region live in a web of insecurities, conditions very often not created by them. Looking deeper into the causes one could conclude that it is certainly a condition which has emerged due to the failure of the state to address the issues effectively surrounding the region. Apparently, the indicators for insecurity in the region are poverty and unemployment for which the fallout is huge. In order to ensure human security, efforts should be placed on an approach which would place its emphasis on developing human capital (no school dropouts; ample scope for skill enhancement; good governance; access to jobs; employment opportunities, etc.).

Often cited by many people that insurgency and militancy are primary reasons for the region's backwardness, many hidden elements are often ignored, such as deficient infrastructure, limited and constrained access to opportunities, and failure of governance. In order to convert the available resources in the region, a long-term strategy should be employed which should ideally reflect on the important aspects of human security in the social, economic, political, environmental and other relevant spheres. Against this backdrop, it is important to index the status of human security and gauge the capacity of the state's norms and institutions to deal with the threats of insecurities.

A composite index of human insecurity would include 14 indicators—poverty, rural youth unemployment, school dropouts, infant mortality, morbidity, crimes against children, crimes against women, incidences of corruption, encroachment of forest area, non-adoption of the right policies, gap between budgeted and actual spending in social sectors, pending cases in courts, occurrence of incidences of violence and presence of security personnel. The index would be helpful in ranking the states according to their level of insecurity within the region as well as in the context of the country.

The indicators can be further divided into three categories such as development, threat and governance, and the states can then be ranked accordingly. In order to develop a rational ranking, factor analysis can be employed, which would provide weights to the indicators based on their dispersion in the data matrix. Certain factors (insecurity of women and children, unemployment, pending cases in courts, poverty and encroachment of forest) rank Tripura as the most insecure state for human beings in the region, followed by the states of Assam and Manipur. In the context of development insecurity, however, Assam ranks first in the region, followed by Tripura and Nagaland. Threat insecurity is perceived to be highest in the state of Tripura, followed by Manipur and Nagaland.

The indicators of governance insecurity show that Meghalaya is the most poorly governed state in the region, followed by Manipur and Assam. Including the country average in the ranking, it is reflected that overall the country ranks higher in the insecurity index than most North-East Indian states, and that governance indicators rank the country highest in the insecurity index. On the other hand, the country as a whole is better placed in terms of insecurity from threats than the seven states of the region. The study indicators largely reflect negative outcomes resulting from an absence of approaches to good governance, as the approaches fail to deliver the dividends of livelihood security and socio-economic prosperity in the region (Acharya and Singhdeo, 2011:4-5).

Over five decades following the Independence of India, it is commonplace to say that the strife-riven states of North-East have been a region which has suffered neglect. In the context of national political consciousness, this perhaps is a correct evaluation. In terms of developmental allocations and a wide range of social and economic indices this is far from reality. The states in the region are ascribed 'special category' status by the Government of India and the National Development Council (NDC) earmarks 30 percent of the total plan allocations to these states as central assistance for those ascribed as 'special category'. Thus, these states receive 90 percent of plan assistance as grant, and 10 percent as loan, as against the norm of 30 percent grant and 70 percent loan for other states. The Finance Commission of India also has a favourable attitude towards these states in terms of the sharing of central tax revenues.

Therefore, clearly these states have not been made to undergo difficult circumstances as a result of their resource endowments. There are a number of other parameters such as the network of roads, industrial development, etc., in which these states lag far behind vis-à-vis the other states in mainland India. But the image of grinding poverty that is often projected is inaccurate, and the 'social backwardness' that we ascribe to the tribes of this region is a heavily concept-dependent value judgment. With lack of productive investment, basic infrastructure facilities and prevalence of unskilled labour force, these states remain economically backward and reflect the character of agrarian economies with two-thirds of its work force dependant on agriculture.

Still the region is not self-sufficient in agricultural production. Industrially, the region is almost marked by non-existence of industries or manufacturing units. Only a few large and medium-scale industries have come up, mostly in Assam, Meghalaya and Arunachal Pradesh. The types of industries that prevail are tea, plywood, jute, petrochemical, paper mill, and fertilizer industries. Some agro-based and forest-based small-scale units have come up and scattered in different states—about half in Assam followed by Tripura and Manipur. The traditional industries in which production is carried out at substance level are handicrafts and weaving industries.

The swelling tertiary sector is the result of an arbitrary arrangement in which a lion's share of states outlay are used in paying salaries and maintaining the large number of unproductive government employees. Naturally, the financial resources of the states including central transfers are directly used to meet consumption needs of the employees that otherwise could have been used for productive investment, escalation of assets, infrastructure, etc., which are so essential for industrial as well as economic growth and overall economic development (Mahajan, 2009:65).

Indeed, the character and equity of local institutions for village and self-governance in tribal societies tend to be far superior to the caste riven, feudal panchayats in much of rural India, the status of women is certainly higher, and crimes against women dramatically lower what they are elsewhere in the country. Our impression of backward or 'primitive' tribal societies in the North-East is cast, at least in some measure, in the mould of the market economy and its indices, and in the patterns of our cultural prejudices (Sahini and George, 2001).

Conclusion

In the context of North-East India, the idea of human security is very interesting as it puts the concept of human development at the forefront of analysis in the region in present times. Although the region has been granted special category status and there is a flow of funds from the centre to each of these states, yet it is a mere eyewash in the context of the region's people. Two perspectives are clearly present in this context: (i) the views of the state and (ii) the views of the local people. The Government of India may boast of allocation of developmental funds with a special focus but on the other hand the allegation that the region has not developed is also true. While the idea of human security is to ensure the freedom from fear and want, the region clearly lacks development and also in the context of the two fundamental aspects of human security. Life in the region is beset with fear of violence and the yearning for peace is ever present. The idea of freedom from want is something which the people of North-East may have a long way to go. 'Want' in the context of North-East arises from lack of development following which the problem of unemployment persists in the region.

Scholars who had worked on human security issues in North-East India conclude with the following points which are a threat to their security and those initiatives which may result in development in this context. In this, they identify that: (i) armed groups are the primary sources of threat, (ii) the state is also a source of threat which arises primarily through counter insurgency measures during the process of maintaining law and order in the region. Based on these local perspectives, many scholars believe that there should be a strengthened emphasis on the conflict resolution process. Issues to be addressed include: (i) insecurity caused by weak governance provides a fertile ground for violence, (ii) the repealing of certain provisions of the AFSPA would result in the reduced power of the security forces and also reduce impunity, and (iii) the local population expect conflicts in the region be resolved through dialogue and not through repression. In the above points, it is evident that human security in the context of North-East India is paralleled with the conflicts resulting in lack of development and infrastructural facilities.

The Indian government may boast of a number of initiatives which look to improve the situation but along with this the government also carries the misconceived idea that any developmental initiative is possible in the region. Such attitudes are sought to be achieved through brute force or through repression. The ground realities and the local population's views and opinions are often ignored and the governmental initiatives are sought to be implemented through force which often attracts violent protests and again the government seeks to suppress these protests through the armed forces that are enabled to act with impunity through legislative measures such as the AFSPA. The categorization of the region as North-East itself presents the attitude of the Indian state in dealing with the problems which beset the eight states. The alienation of the region by such attitudes is ever present since the integration of the region with the Indian union.

Thus, in conclusion, it should be noted that human security in the region is essentially a significant and important aspect for development of the region. However, such initiatives should be undertaken by taking into account the local population's needs and opinions which clearly are important to solve the potential future possibilities of violent conflicts in the region.

Endnotes

1. Foreign ministry website, available at: http://www.dfait-maeci.gc.ca/foreignp/humansecurity/menu-e.asp.
2. Government of Japan, Ministry of Foreign Affairs, "Diplomatic Bluebook 1999", Chapter 2, Section 3, available at: http://www.mofa.go.jp.
3. The theory of youth bulge can be applied in this context as the claim youth bulges cause violent conflicts have a long history, it has received more attention following the September 11, 2001 for explaining current political instability (Urdal, 2004:1). The available literature on youth bulge focuses on spontaneous and low intensity unrest like non-violent protest and rioting. However, it also suggests that youth bulges may increase the likelihood of more organized forms of political violence like internal armed conflict (Urdal, 2012:1). Violent conflict as a result of youth bulge can be explained using two dominant and competing theoretical traditions, one focusing on opportunity and the other on motive. Both perspectives are macro-level frameworks which look to explain decisions of individuals to join or participate in violent conflict by focusing on economic,

political and social factors. Often described as the greed perspective and having its roots in micro economic theory, rebellions might arise as rebels aspire to capture wealth and resources extra-legally (Collier and Hoeffler, 2004:564). According to this perspective, individuals will opt to join rebel groups only when the potential gain by joining the groups yields high returns in comparison with expected costs (Collier, 2000: 94). If economic agents are actuated by self-interest, then it must be demonstrated why individuals choose war over other alternatives for income generation. Thus, theorizing the greed perspective must be based on economic motivations for violence and criminality. Such groups in general do not show any interest in preserving the state or its people but are simply with the intent of loot (Murshed and Tadjoeddin, 2009:90).

On the other hand, the motive-oriented or the grievance perspective is drawn from the relative deprivation theory and tends to analyse the outbreak of political violence as a means to redress economic or political grievances (Gurr, 1970:223). Much of the literature focuses on this perspective of grievance where lack of political will to address issues faced by large youth cohorts crowding the labour market and agglomerated in urban centres, pave the way to political violence (Choucri, 1974:83; Braungart, 1984:7). Civil wars which happen in the context of grievances are mainly where societies are multi-ethnic or multi-religious. Here political exclusion or victimization of particular groups can occur more particularly against groups which are a minority-ethnic dominance (Collier and Hoeffler, 2002:13).

References

Acharya, Amitav et al. (2011), "Human Security: From Concept to Practice–Case Studies from North-East India and Orissa", Singapore: World Scientific Publishing.

Baruah, Sanjib (2002), "Gullivers Troubles: State and Militancy in North East India", *Economic and Political Weekly*, 37(41): 4178-4182.

Bhaumik, Subir (2007), "Insurgencies in India's North-East: Conflict, Co-option and Change", Washington D.C.: East West Centre.

Boyle, Kevin and Simonsen, Sigmund (2004), "Human Security, Human Rights and Disarmament", Disarmament Forum,

Geneva: UNIDIR.

Chen, Lincoln (1995), "Human Security: Concepts and Approaches", in Tatsuro Matsumae and Lincoln Chen (eds.), 'Common Security in Asia: New Concepts of Human Security', Tokyo: Tokai University Press.

Commission on Human Security (2003), "Human Security Now", New York: Commission on Human Security.

Das, Samir Kumar (2007), "Conflict and Peace in India's North-East: The Role of Civil Society", Washington D.C.: East West Centre.

Dev, Rajesh (2004), "Human Rights, Relativism and Minorities in North-East India", *Economic and Political Weekly*, 39(43): 4747-4752.

Gasper, Desmond R. (2005), "Securing Human Security: Situating 'Human Security' as Concept and Discourse", *Journal of Human Development*, 6(2): 221-245.

Gill, K.P.S. (1998), "Injustice of Campaign Against Police", *The Pioneer*, October 17.

Ginkel, Hans Van and Newman, Edward (2000), "In Quest of Human Security", *Japan Review of International Affairs*, 14(1): 59-82.

Gomez, Oscar A. and Gasper, Desmond R. (2013), "Human Security: A Thematic Guidance Note for Regional and National Human Development Report Teams", available at: hdr.undp.org/sites/default/human_security_guidance_note_r-hdrs.pdf.

Grayson, Kyle (2004), "A Challenge to the Power over Knowledge of Traditional Security", *Security Dialogue*, 35(3): 357.

Hammerstad, Anne (2000), "Whose Security? UNHCR, Refugee Protection and State Security after the Cold War", *Security Dialogue*, 31(4), 391-403.

Hussein, Karim, Gnisci, Donata and Wanjiru, Julia (2004), "Security and Human Security: An Overview of Concepts and Limitations, What Implications for West Africa", Paris: OECD.

King, Gary and Murray Christopher J.L. (2002), "Rethinking Human Security", *Political Science Quarterly*, 116(4): 585-610.

Mahajan, Samir K. (2009), "Attainment of Human Development: A Study of North-East India", *Delhi Business Review*, 10(2): 63-71.

Obleitner, Gerd (2002), "New Threats to Global Security: Challenges for the United Nations and Human Multilateralism", Paper

Presenter at the 2002 Annual Meeting of the Academic Council on the United Nations System, Graz: European Training and Research Centre for Human Rights.

Paris, Roland (2001), "Human Security: Paradigm Shift or Hot Air?", *International Security* 26(2): 87-102.

Rajagopalan, Swarna (2008), "Peace Accords in North-East India: Journey over Milestones", Washington D.C.: East West Centre.

Tadjbakhsh, Shahrbanou and Chenoy, Anuradha M. (2007), "Human Security: Concepts and Implications", New York: Routledge.

Tadjbakhsh, Shahrbanou (2009), "Human Security: Looking Back Before Looking Forward", Paper for the ICHSWA Conference, Birjand, November 26, 2009, available at: www.birjand.ac.ir/ichwsa/downloads/Dr%20Tadjbakhsh%275%20Paper%ICHWSA.pdf.

Thakur, Ramesh (1997), "From National to Human Security", in Stuart Harris and Andrew Mack (eds.), 'Asia-Pacific Security: The Economics-Politics Nexus', Sydney: Allen and Unwin.

6

Natural Disasters and Relocation in India

Rajeev M.M.

Introduction

The developing countries are becoming increasingly exposed to greater numbers of natural and manmade disasters, resulting in larger numbers of people becoming victims. The social and economic costs of disasters vary widely and are difficult to estimate on a global basis. From 1980 to 1984, about 800 disasters affected the lives of about 400 million people in the world and during the period 2000 to 2004, about 1,300 disasters affected the lives of about 1.4 billion people, which means there is an increase in the number of disasters and in the number of people affected during the past few decades. According to the Bureau of Crisis Prevention and Recovery (BCPR) of the United Nations Development Program (UNDP), around 75 percent of the world's population live in areas that have been affected by natural disaster once in a year.

On December 26 in 2004, an earthquake in the Indian Ocean caused a tsunami that damaged 4,500 km. (2,812 miles) of coastline. Out of the countries impacted, Indonesia was the worst affected as it was the closest to the epicentre (Gray et al., 2009a; Gray et al., 2009b). This study found that the patterns of mobility is related to the degree of damage; in severely affected areas, two out of three adults (15 years and older) moved, compared to one in eight in lightly damaged areas. Similarities between the Indonesian and Japanese relocation are that people in Japan moved beyond jurisdictional boundaries and the primary refuge was at a private residence compared to a hotel/inn or shelter. It should be noted that mobility patterns differ across geography and types of natural disaster. For example, in Bangladesh, crop failure has a strong impact on the mobility of those living in severely affected areas but flooding has pronounced impact on women and the poor (Gray and Mueller, 2012).

India has been traditionally vulnerable to natural disasters on account of its unique geo-climatic conditions. The Indian disaster profile shows that 8 percent land is prone to cyclone, 12 percent of total land mass is prone to flood, 28 percent of the land is vulnerable to drought and 56 percent of the land mass is prone to earthquakes. India has experienced many major and minor disasters in which the most devastated disasters were the super cyclone of Orissa (1999), the Gujarat earthquake (2001), and Tsunami (2004) which affected millions across the country and resulted in heavy loss of life, property and livelihood (UNDP DMTP, 1992).

There were many manmade disasters in India which might have been intentional or due to the negligence from the concerned authorities, which included the Bhopal gas tragedy and the likes. All these resulted in huge loss of life, livelihood and property. The impact of disasters touches various areas including physical, psychological, social and economical. The impact on each factor will depend individually on their ability to cope up with their own capacity of resources which they possess. The poor are the ones most vulnerable to disasters because they have fewer resources and capacity to prevent or cope with the disasters and the impacts affect the quality of life of the victims.

Tsunamis are among the most terrifying natural hazards known to man and have been responsible for tremendous loss of life and property throughout the history. Because of their destructiveness, tsunamis have/had notable impact on the human, social and economic sectors of our societies. Tsunami has destroyed large coastal population. Destructive tsunamis have also occurred in the Indian Ocean and in the

Mediterranean Sea (Department of Ocean Development, 2005). The housing stock along the coast, as well as bridges and roads, suffered extensive damage. Structures were damaged by direct pressure from tsunami waves, and scouring damage was induced by the receding waves. Many of the affected structures consisted of non-engineered, poorly constructed houses belonging to the fishing community.

Relocation should be treated as a last resort. A review of tropical cyclone response in the Philippines from 2009 to 2011 found that 'relocation of affected people outside of their areas of origins should be the last option (Cosgrave, J., 2014). The World Bank opined, "Avoid relocation if at all possible. Especially avoid relocation to distant sites" (Cosgrave, J., 2014). A review of five years after the 2004 Indian Ocean tsunami revealed that 'most relocated communities are much further away from important facilities than before the tsunami and suffer from higher costs of transport and less job opportunities" (Cosgrave, J., 2014). Such relocation can delay communities' development. Relocation may move people away not only from their livelihoods but also from their kinship networks and social support systems, as what was witnessed in India after the 2004 Indian Ocean tsunami. Too often, agencies regard relocation as simply a matter of building houses. The World Bank's Handbook for Reconstruction after natural disasters made the point that 'relocation is not only about re-housing people, but also about reviving livelihoods and rebuilding the community, the environment, and social capital" (Cosgrave, J., 2014).

It must be recognized that the relocation of communities is a complex task that requires a holistic approach which is consultative in nature and led by relevant national and local authorities. It should only take place as a last resort after all options have been exhausted. It is important to ensure that states and supporting actors undertake relocation in a manner that respects the human rights of affected populations. Obtaining the free and informed consent of affected communities (both relocating and host communities) requires their consultation and participation in the planning and decision-making process of relocation. At the international and regional level, disaster risk management practitioners engaged in human mobility—such as through management of planned relocation, migration and displacement—need to come together with national authorities to ensure that planned relocation efforts are integrated into strategies related to disaster risk reduction strategies and climate change adaptation efforts. International stakeholders should especially work to facilitate identification and dissemination of best practices, guidelines and methodologies for successful relocation.

Poor and vulnerable populations often face permanent relocation after a disaster, conflict or a large infrastructure development project such as construction of a dam. Even short-term displacement is disruptive, but resettlement can tear apart communities and leave families without the assets they need for sustainable livelihoods (Kathryn Reid, 2014). Major post-tsunami studies have concluded that the immediate relief effort was largely effective but the long-term social recovery work has been less impressive. Furthermore, the post-tsunami rehabilitation work involved the most extensive relocation and resettlement of dislocated people yet undertaken and so it is critical to learn important lessons from this resettlement process. In the days immediately after the disaster, they were given, or found, shelter away from their beaches, often in temporary refugee camps. Many wanted to return to the areas where they used to live so that they could start rebuilding their homes and livelihoods.

The Oxford dictionary defines relocation as "move to a new place to live" (Oxford Dictionary, 1995). Experiences have showed that improved housing provided by habitat was linked to health benefits and a higher standard of living. Overall, habitat's community-based disaster management approach, with its strong focus on partner-family participation and fairness in family selection, was seen as enhancing community cohesion and cooperation. These findings were true of relocated communities as well as those that were able to rebuild on location (Kathryn Reid, 2014).

Looking back, the Indian government took the opportunity to enforce and ensure the compliance of building codes and introduced a scheme to provide houses to people who were willing to stay 200 mts. away from the sea. Government launched a policy that permanent houses within 200 mts. from the coastlines will principally be relocated to safer places. Damaged houses reconstructed on the land acquired by the governments beyond 200 mts. from the coastlines are provided subsidy by the NGOs or governments. The other victims who did not have any shelter were provided relocation with the help of various agencies and NGOs.

Relocation is defined as a process whereby a community's housing, assets, and public infrastructure are rebuilt in another location. Relocation is sometimes perceived to be the best option after a disaster for one or more of the following reasons: (i) people have already been displaced by the disaster, (ii) their current location is judged to be uninhabitable, or (iii) relocation is considered the best option to reduce vulnerability to the risk of future disasters.

In post-disaster settings such as those found in the aftermath of the tsunami, the urgency and scale of the need for shelter poses a tremendous challenge. Reconstruction also looks beyond the structure of a single house to how settlements are laid out. The designs of houses, infrastructure and green spaces of a neighbourhood shape the larger settlement and, therefore, have a larger effect on the community's and the user's energy use, biodiversity and quality of life. The relocation to distant places from the native place will negatively affect the physical, psychological, social and environmental qualities of life. The studies have explored reasons for people displacement, various types of relocations, the relocation process, and also how to improve the quality of life of re-settlers. In turn, they have come up with useful concepts, analytical approaches and models that have broadened our knowledge and understanding of relocation as a social phenomenon on the whole (Oliver-Smith, 1990; Cernea and McDowell, 2000; Muggah, 2008).

The relocation of the population from their habitat to other remote or unknown areas made them vulnerable and resulted in poor quality of life. The quality of life means an overall well-being of an individual. Also, it is fundamental to balance the factors which contribute to the quality of life. The present study thus focuses on the "quality of life among the tsunami affected relocated and non-relocated families—6 years after the disaster (tsunami). The researcher values the study as very significant in terms of understanding the different factors incorporated in the quality of life of disaster survivors in the two unique contexts of relocation and non-relocation.

Disaster relocation may be appropriate when the disaster is the result of site-specific vulnerabilities. One of the chief reasons for relocation failure is underweighing the welfare of the population as a criterion for the selection of the relocation site. Inappropriate land may be chosen for a relocation project because it can be acquired quickly, is owned or controlled by government, or is easily accessible with topography that favours rapid construction. For similar reasons, people resettled to protect them from one risk (e.g., tsunamis) may find themselves exposed to new ones (e.g., risks to livelihood, high crime, lack of services). Relocation of communities requires risk mitigation through well-planned and adequately financed programs that include such elements as land-for-land exchange, employment generation, ensured food security, improved access to health services, and transportation to jobs (Oliver-Smith, 1990;1991a; 1991b; 1996; 2005; Scudder, 1981; 2005; 2000; Muggah, 2008; De Wet, 2006).

Importantly, these studies have explored reasons for people displacement, various types of relocations, the relocation process, and also how to improve the quality of life of resettlers. In turn, they have come up with useful concepts, analytical approaches and models that have broadened our knowledge and understanding of relocation as a social phenomenon as a whole. From all these studies, it is understood that relocation may affect the quality of life of tsunami affected people either due to wrong site selection or remote areas. All may affect the quality of life and the related factors.

An effective relocation plan is one that the affected population views positively. Relocation is not a decision; risk may be sufficiently reduced simply by reducing the population of a settlement, rather than by relocating it entirely. Relocation is not only about re-housing people, but also about reviving livelihoods and rebuilding the community, the environment, and the social capital. It is better to create incentives that encourage people to relocate than to force them to leave. Relocation should take place as close to the original habitat as possible. The host community is part of the affected population and should be involved in planning (GFDRR, World Bank, 2010).

The main aim and objectives of the present were: (a) to understand the problems and challenges encountered by the families in the relocated villages; (b) to find out various activities performed by different actors during the time of relocation of disaster survivors; and (c) to suggest appropriate strategies and methods to be adopted during relocation.

The investigators studied 30 tsunami affected survivors as samples through purposive sampling method. For the qualitative research, an interview guide was prepared by the investigator. The interview guide contained questions specifically related to the problems encountered by the people in managing the disaster situation. The qualitative interviews were conducted in Malayalam by the principal investigator and lasted for one to one-and-a-half hours. All the interviews were audio recorded and later transcribed for analysis.

Major Findings

The participants in the study were concerned about the following problems in the relocated villages:

"Fish on Land": Participants used the "fish on land" metaphor to describe their situation in the relocated villages. In the relocated neighbourhoods, clear division existed between families in the newly formed village. Some behaviours that were normal in their old neighbourhood became kind of a concern for the newly formed communities. For instance, in their old neighbourhood, it was perfectly alright for people to ample through the land, backyard or even frontyard of their neighbours. It was culturally approved and had implicit community sanction. However, in the relocated villages, this behaviour was considered as trespassing and was very often a reason for conflict between the neighbours. In addition, relocated families frequently got stopped by others while they went for fishing at night. Furthermore, families felt like being strangers in the relocated places.

Lack of Cultural Consideration while Relocating: Study participants felt that the relocation process was haphazardly planned and did not take into consideration the cultural characteristics and employment prospects of the relocated individuals. Government officials did not consult disaster survivors when they acquired the land for relocation. The acquired relocation land was far from the coastal areas making it difficult for fishermen to access the sea, particularly during night because of the transportation issues. This had a negative impact on individuals and families as they depended solely on their fishing income.

Tsunami Colony: People started labelling the newly formed communities as "tsunami colonies". People settled in the new communities were considered as refugees in need of special help and treatment by the neighbouring villages. People felt stigmatized and isolated in these colonies and this was a growing concern for many of the families.

Poor Services at the Relocated Villages: Families felt that there were poor water and sanitation services in the relocated villages. In some cases, there wasn't even a desirable place to perform the funeral rites and many families were cramped together in a small area. The local politicians initially adopted a distant attitude towards the newcomers which escalated their problems. Participants felt that had they been involved in the relocation process and had there been a more proactive role from the LSG, it would have helped to ameliorate some of their sufferings.

Other significant findings were: (i) relocation process should have engaged NGOs and aligned their priorities with larger reconstruction programmes; (ii) importance of clarity in public policies regarding relocation and occupation of environmentally sensitive areas should have been accorded priority such as buffer zones etc.; and (iii) the livelihood and housing reconstruction requirements of the same population in a larger way should have been addressed.

Recommendations

The researcher makes the following recommendations:

1. Necessities which are to be considered before the relocation process like information about the relocation (when, where, for how many) should be properly looked into.
2. Site of selection should be done after local consultations.
3. Relocation site should have easy access to medical services, especially services for pregnant and lactating mothers and children.
4. Relocation should take into consideration women's livelihood needs, markets, and other local facilities.
5. Physical security of women and children in relocation especially women who have lost their families, should be ensured.

The current study emphasises the need for community based sensitization programmes on disaster relocation and related issues to resolve the post-disaster difficulties. For successful relocation, site of the relocation, community ownership and local governmental support should ensure effective and efficient implementation to promote long-term support and quality care for the disaster affected people. The study concludes by highlighting the significance of further studies and actions in the areas of disaster relocation, resettlement, education, long-term support and quality of life of survivors of any kind of disaster and related context.

Conclusion

Relocation of disaster victims is always a challenging task for agencies involved in disaster reconstruction. It is important to note that relocation of disaster victims is more than moving a group of people from one place to another. To build a strong community is important for people moving from different areas into the same place. If affected people are to build a permanent home after the disaster, developing strong community solidarity is important. It is vital for agencies involved in post-disaster reconstruction to recognize this issue and help survivors build a strong community.

References

Bureau for Crisis Prevention and Recovery-UNDP (BCPR-UNDP), 2013.

Cosgrave, J. (2014), “Responding to Flood Disasters: Learning from Previous Relief and Recovery Operations”, ALNAP Lessons Paper, London: ALNAP/ODI.

Department of Ocean Development (2005), “Preliminary Assessment of Impact of Tsunami in Selected Coastal Areas of India”.

Gray et al. (2009a; 2009b), “Disaster and Relocation, Migration Case Study: How Displaced People Respond and Adapt to Natural Disasters?”, AAG Centre for Global Geography Education.

International Food Policy Research Institute (IFPRI).

Kathryn Reid (2014), “Post-disaster Relocation Poses Extra Challenges”, Habitat for Humanity International.

Oxford Dictionary (1995).

“Safer Homes, Stronger Communities: A Handbook for Reconstructing after Natural Disasters (2010)”, The International Bank for Reconstruction and Development, World Bank, Washington D.C.

UNDP (1992), “An Overview of Disaster Management”, UNDP-

UNDRO Disaster Management Training Programme Manual, 2nd edition.

United Nations Environment Programme and SKAT (2007).

7

Environmental Governance and Conservation of Western Ghats

Anju Liz Kurian and C. Vinodan

Introduction

Rising consequences and atrociousness of environmental challenges are restricting the prospect of economic development especially in poor countries and developing regions of the world. A number of studies and an extensive assessment about the current status of the environment remind us that the use of our planet's resources and environmental services being provided are far being outstripped and the planetary boundaries are being close to tipping if continued in business in the usual scenario (Singh, 2014:601). Governments are persisting on setting up international mechanisms for addressing global scale environmental issues which has led to a labyrinthine international bureaucracy. As a result, it turns to create significant burdens on national administrative capacity in both the developed and the developing world and most importantly, inability on the part of existing international or national bodies to effectively address the problems in sight.

In this context, the question of the most appropriate governance architecture for the scale and scope of contemporary global environmental problems has become an important focus in both policy and academic debates. Scholars and politicians alike have argued that if we do not address the governance failures, our stewardship of the environment will continue to be ineffective and inequitable, with little chance of finding a path towards sustainability (Ivanova, 2007:4).

At the international level, there is a remarkable progress in terms of policies and creation of environmental authorities for governance and environmental commitments for conservation, particularly in the scenario of global warming and sustainable development. Yet, there is a widening gap between the environmental commitments made and the actual implementation to recover environmental consequences. In most of the circumstances, environmental policy design is embedded in a political context with multifaceted drivers. Actually, the instruments that reinforce significant human rights principles, such as the rule of law, transparency and public participation, may be equal or more imperative than specific environmental policies or projects so as to advance environmental governance.

It is ever more acknowledged that technical solutions to environmental problems are not adequate in achieving sustainable development. On the contrary there needs an increasing attention to the value of governance to manage the wide range of environmental challenges and impacts. The attention given to environmental conservation in the backdrop of global warming and climate change opens new avenues of opportunities and risks for environmental governance (Slunge and Emelie, 2010:19).

India has a robust history of nature conservation, going back to pre-historical times. These traditions follow many norms that are being recommended today such as the criteria for ecologically sensitive areas (ESAs), sacred forests protecting origins of rivers for instance, Bhimashankar in Pune District, Maharashtra, at the origin of the Bhima River or important breeding habitats for example, Kokkre-Bellur Pelicanry in Mandya District, Karnataka etc. Nowadays, there have been identified and established substantial set of conservation sites namely biosphere reserves, national parks and wildlife sanctuaries which represent a fairly successful network of protected areas (PAs) for conserving the stunning biodiversity. Generally, most of these areas are large forested lands and they harbour high levels of biological diversity, flagship species, or unique landscape elements (Gadgil et al.,

2011:177).

In reality, India's first wildlife protected areas and forest reserves were not set up by the government, or by kings or sultans. They were created and well nurtured by ordinary people who set aside parts of the landscape for cultural, ethical, or economic reasons. Several instances of conservation efforts of natural ecosystems and wildlife populations by local communities are well documented in India. A number of sites conserved by communities have been recognized to be of such wildlife value that they have been declared wildlife sanctuaries or national parks by the state governments (Kothari and Neema, 2010:182-185).

Environmental Governance and Conservation

Environmental governance is a precise form of the broader term 'governance' which refers to the processes and institutions through which societies formulate decisions that impinge on the environment *via* a normative dimension of sustainability. In short, environmental governance is mainly aimed at attaining environmental conservation goals sustainably which can be cyphered by the effectiveness of strategies as well as initiatives implemented to accomplish the objectives. Prior informed consent of the stakeholders is the cornerstone of conservation agenda irrespective of geographic realm. Stakeholder access to information, adequate funding, transparency and accountability, are other critical aspects of proper environmental governance. Global environmental governance can be perceived as the organisations, policy instruments, financing mechanisms, rules, procedures and norms, which control the method of global environmental protection (Wingqvist et al., 2012:14-15).

Therefore, environmental governance can be outlined as, 'the ability of a state to govern its resources as prescribed in forms of legal instruments and supplemented by policy, programmes and institutional interventions, all operating in a holistic manner synergistically among and within various entities, taking into consideration the social, cultural, environmental and economic factors' (Singh, 2014:604). Generally, environmental governance is formed by a set of multilateral environmental agreements (MEAs), political agreements, non-binding agreements, programmes, projects and national laws executed at different levels. Many policies in connection with the environment are determined at the international level while others are done at the regional and sub-regional levels which are endorsed by a multitude of stakeholders and a number of implementing agencies (Singh, 2014:601). Even though numerous multilateral agreements have been signed and ratified, its implementation poses severe problems at the national, regional and international level whilst environmental degradation is increasing steadily (Singh, 2014:607).

India is a signatory to various international conventions that affect forest and biodiversity conservation—Convention on International Trade in Endangered Species of Wild Flora and Fauna (CITES), 1973, Convention on Wetlands of International Importance especially as Waterfowl Habitat, United Nations Framework Convention on Climate Change (UNFCCC), 1992, Convention on Biological Diversity (CBD), 1992, United Nations Convention to Combat Desertification (UNCCD), 1994, Convention on the Conservation of Migratory Species of Wild Animals, Bonn, 1979, Convention for the Protection of the World Cultural and Natural Heritage, 1972, International Convention for the Regulation of Whaling, 1946, and the United Nations Convention on the Laws of the Sea (UNCLOS), 1982.

India has been committed to the implementation of Chapter XI of Agenda 21 and the non-binding Forestry Principles and has developed appropriate national strategies, legislations and administrative instruments to address the indebtednesses under the conventions. India believes that all concomitant weightings on the international arrangements or mechanisms should be geared by and be in consonance with the elements for action identified in Agenda 21 and the Forestry Principles. Together with the Convention on Biodiversity (CBD), the Land Use, Land Use

Change and Forestry (LULUCF) sector under the Kyoto Protocol and Paris Agreement of the United Nations Framework Convention on Climate Change (UNFCC), the country is obliged internationally to urge for ecological security (UNEP, 2005; UNFCC, 2015).

Effective institutions at national level, with capacities to implement national environmental legislation, can promote and support the intents of the MEAs. Strong environmental authorities are trustworthy for implementing MEAs at the national level in true spirit. Nevertheless, environment ministries in developing and transitional countries are typically weak as they commonly have difficulties in getting resources from the treasury/budgetary allocation, while international finance for environmental projects was more charismatic. The project oriented approach was often blind towards the root causes and drivers to environmental degradation which may even sabotage the influence of environmental authorities. There is hence a risk that environmental agencies spend a disproportionate share of their scarce human and administrative resources on negotiating and administering internationally financed projects instead of focusing on executing their core functions such as monitoring, control and supervision (Lawson and Neil, 2008).

The fragmentation and projections of the international environmental governance system is mirrored in the implementation of MEAs at the national level. National action plans and associated projects are the main vehicles for translating MEAs into practical action at the national level. However, these action plans have been criticised for being too project focused and poorly integrated with national development planning. Without a comprehensive strategy, opportunities are often lost to capitalise on potential synergies between different environmental measures. Despite the emphasis on consultation and national ownership in the development of these plans, many action plans seem to have been primarily developed for the purpose of attracting international funding for different projects (Sharma, 2009:7).

It should be noted that the true potential of national action planning under MEAs lies in their climacteric role in offering active participation by the stakeholders in global and national environmental governance and enabling countries to take a coherent and strategic approach in integrating environmental concerns into national development priorities.

Environment lobbies such as government environment departments, government and non-government organisations and scientific institutions will have to shift their focus from winning over the support of donors to winning support of national governments and stakeholders by convincing them of the wider benefits of implementing relevant international agreements (Sharma, 2009:40, 41) without interfering or affecting stakeholder livelihood issues and existence. Even though addressing environmental challenges is the need of the hour, the associated large flows of funds that are envisaged as a response to these challenges makes the situation more complex. The cornucopia of funding opportunities provides additional arguments for the need of good governance while its implications on stakeholders should not be ignored in the current era of globalization and environmental colonization.

The plethora of multilateral and bilateral funds and private sector finance and a variety of financing instruments like grants, loans, guarantees, technology transfer, etc. to combat global warming is leading the recipient countries into increasing fragmentation and unnecessarily high administrative and institutional encumbrance (Thornton, 2010). The external, project based, fund recipients among the majority of environmental authorities paves the way for challenges in environmental governance in developing countries. This leads to several linked issues and risks such as most of the project proposals are based on international rather than national priorities, authorities are aimed at project management rather than strategic governance which naturally tends to by-pass country systems and their accountability is mainly to external financiers rather than to citizens (OECD, 2012:22).

With the support of donors, lion part of environmental organisations have successfully initiated processes of preparing environmental policies, environmental plans such as National Environmental Action Plans, National Conservation Strategies, etc., framework environmental laws, environmental impact assessment (EIA) guidelines and procedures and environmental standards, environmental databases and numerous environment related publications (OECD, 1999:16) with a myriad of repercussions on local community as well as the society.

Protected areas (PAs) intended specifically for biodiversity conservation in India continue to be extremely preservationist in character. PAs networks are governed by staggeringly rigid management regimes with local communities having little role in utilization, management and conservation. This approach has led to several difficulties, as assessed comprehensively by the Tiger Task Force. It should be noted that the nature conservation regime must also now take into account new legislation such as the Scheduled Tribes and Other Traditional Forest Dwellers (Rights over the Forests) Act, 2006. Furthermore, modern ecological theory stipulates strongly an adaptive co-management approach where a network of conservation sites that have locality and time-specific, flexible management strategies. ESAs can be identified with such a flexible system of management. In fact, there could be ESAs with PAs embedded within them, with an adaptive regime of regulation (Gadgil et al., 2011:178).

Globally protected areas have expanded threefold recently, and the stricter category 1-4 protected areas now number about 49,000 and cover six percent of the land surface of the planet. In the case of India, the number of PAs has increased thirteen-fold in the last three to four decades with some evictions. A consequential resettlement of more than 100,000 people and denial of traditional lands in many circumstances has resulted in this regard. Moreover, this has affected millions of people whose livelihoods depend on non-timber forest products

(NTFP), which contribute between 14 to 23 percent of the rural income for economically weaker communities who are marginalized in the society.

The International Union for Conservation of Nature (IUCN) in 1994 classified the world's PA systems into six categories, two of which (Category V and Category VI) tried to integrate biodiversity conservation and the resource needs of the local people (Kanagavel et al., 2013:61). The steady increase in human-wildlife conflicts is a matter of serious concern with multitudinous reverberations across the society. Interestingly, in many places, the regeneration of ecosystems has increased wildlife populations with spilling out phenomenon and intermittent crop raids, threats to livestock or even human life. Man-animal conflict arena is comparatively new which is steadily increasing especially in developing countries where livelihood and conservation agendas are competing for survival as well as existence. Communities have been seeking help to resolve these conflicts, which in some places are weakening the resolve to continue conservation efforts (Kothari and Neema, 2010:187).

There is an extraordinary dearth of good information about the social impacts of protected areas. Most of the existing literature merely mentioned conservation displacement or evictions, with none of the details regarding social discontent and livelihood issues of affected population. The importance of this issue is recognized in the Convention of Biodiversity calling for an assessment of 'the economic and socio-cultural costs and impacts arising from the establishment and maintenance of protected areas'. The IUCN Secretary General observed in a meeting at the 2004 World Conservation Congress that a proposed global assessment of the social impacts of conservation was unlikely to receive support within IUCN programs because powerful groups within the IUCN feared that 'the results might be used against conservation itself'.

As legislation and enforcement tightens over protected

areas, policy as well as new areas with conservation interest will lead to eviction of millions of people and creation of conservation refugees. Thus, one of conservationists' austerest forthcoming challenges is how to deal with this auspicating displacement to be carried out in the name of nature (Brockington et al., 2006:250). Conservation activities involve displacement of communities and denial of allied livelihood of the inhabitants which has a global impact. Denunciation of free and prior informed consent by indigenous people and other local communities could further exacerbate these impacts, with respect to their rights, lands and territories. It should be noted that the human rights attributes of such eviction measures need to be carefully considered in designing these policies and projects, irrespective of the geographic region (CIEL, 2011:9).

The conservation efforts are successful if there is an increasing understanding that biodiversity conservation requires the involvement of indigenous communities and local people that live in natural ecosystems. India is also not an exception in the conservation portfolio as there have been a number of policy processes and pronouncements that have stressed the need for broad-based conservation efforts which encompasses the well-being of local communities who depend on natural resources (Kothari and Neema, 2010:181).

Conservation of Western Ghats

Western Ghats (WG) has highest human population density and anthropogenic pressure in the world which needs commendable conservation attention at the international level (Kanagavel et al., 2013:61). The WG extending along the west coast of India covers an area of 160,000 square kilometres and is one of the 34 global biodiversity hotspots for conservation. The Western Ghats, also known as the Sahyadri Hills stretch for 1,600 kilometres along the west coast of India, interrupted only by the 30 kilometres long Palghat Gap, through the states of Kerala, Tamil Nadu, Karnataka, Goa, Maharashtra and Gujarat. WG is famous for its stunning richness in biodiversity.

In addition, the Western Ghats is home to diverse social, religious, and linguistic groups who are the real stakeholders and beneficiaries. There is high cultural diversity of rituals, customs, and lifestyles in the region including a significant population of *adivasis* and forest dwellers. Even though the region is a biodiversity hotspot it supports about 260 people/km^2 and depends on land for agriculture, plantations and perennial crops as well as forms of development, industrialization, mining and tourism (Equations, 2011:5).

The Western Ghats is homage for unique assemblages of plant and animal communities and endemic species is globally important for conserving representative areas of the Earth's biodiversity. The biodiversity coexists with an unusually high density of human population, a significant percentage of which has great veneration for nature. The high cultural diversity of rituals, customs, and lifestyles has led to the establishment of several religious institutions that strongly influence public opinion and the political decision-making process in the region. Thus, conservation in the Western Ghats landscape occurs within a highly diverse and complex institutional landscape. Civil society institutions have a predominant role in influencing conservation agendas in the region. The most powerful institutions that control land use through land ownership include the State Forest Departments and associated development corporations, Government institutions such as the Public Works, Electricity, Irrigation and Revenue departments, private plantation companies, and individual landowners controlling the use of large tracts of land including forests within the Western Ghats. Creative engagement and liaison of these institutions is decisive to achieve effective conservation outcomes across the hotspot (Bawa et al., 2007).

The environmental governance and conservation issues are complexly interconnected with human habitation and livelihood among stakeholders of Western Ghats. Illegal hunting, livestock grazing, and forest fires, and landscape-level threats such as mining, roads, large and micro-hydel power projects, wind farms, large-scale agricultural expansion, and creation of monoculture plantations are the major threats according to hard core conservationists which either independently or synergistically influence biodiversity of the hotspot. However, there is growing concern that the advocates of conservation in WG has failed to address issues of the stakeholders as well as settler community across the WG.

Interestingly, threats are elaborately looped together complexly in myriad ways making it a difficult dispute to tease apart the impingements. The biodiversity of the Hill agro-ecosystems in the Western Ghats are prevailed by tea, coffee, rubber, and monocultures of various species including exotic species like eucalyptus, wattle etc. Traditional farmers in the WG are the real conservationists and any conservation efforts without the participation of local people are counterproductive. None of the hardcore conservationists are ready to address the issues of the stakeholder community across WG which is a serious pitfall and can result in a conservation dilemma among the WG settlers.

The Government of India has submitted nomination dossier of WG for inscription in UNESCO's (United Nations Educational, Scientific and Cultural Organization) World Heritage List which is inscribed at the 36th Session of the World Heritage Committee in Saint Petersburg, Russian Federation (June 24-July 6, 2012). The nomination dossier for WG was prepared by Wildlife Institute of India (WII), Dehradun and two civil society organizations, Ashoka Trust for Research in Ecology and Environment, Bangalore (ATREE), and Nature Conservation Foundation (NCF), Mysore. But the 2006 nomination was rejected by the World Heritage Committee (Bijoy, 2014:224).

Then in 2009, the Government of India resubmitted the nomination to UNESCO for the Western Ghats to be listed as a 'serial' natural World Heritage site. The nomination was prepared based on criterion (vii) which contains superlative natural phenomena or areas of exceptional natural beauty and aesthetic importance and criterion (x) that contains the most important and significant natural habitats for *in-situ* conservation of biological diversity, including those containing threatened species of outstanding universal value from the point of view of science or conservation of the World Heritage Convention's Operational Guidelines.

Ministry of Environment and Forest (MoEF) constituted a Western Ghats Natural Heritage Management Committee on August 31, 2010 for the purpose of 'dealing with matters relating to the inscription and management of the Western Ghats Serial Sites'. They reviewed the itinerary and logistics for the two-member IUCN (International Union for Conservation of Nature) team which visited the sites to assess the scientific, technical and administrative aspects of the proposal through site visits and interactions with scientists, conservationists and government officials. The IUCN team has visited and evaluated the component sites. However, the UNESCO officials were confronted with public ire by various sections of the local population due to the concealment maintained by the forest officials and conservationists around the team's visit. This unilateral and undemocratic move by the forest officials and conservationists makes the situation more complex and arouse suspicion that the whole exercise had an ominous objective of depriving the local inhabitants of whatever little rights they had, and that local inhabitants would be displaced or evicted as a result of the World Heritage designation (Bijoy, 2014:233). Places like Coorg in Karnataka have documented high resistance and public ire against the UN officials (Anonymous, 2010).

After the strenuous exercise by the expert team of the UNESCO across the WG, they requested further clarifications about management regime for the WG in 2010. Meanwhile, the Western Ghats Ecology Expert Panel (WGEEP) was set up by MoEF in 2010 to 'assess the current status of ecology of the Western Ghats region', 'demarcate areas which need to be notified as ecologically sensitive' and 'make recommendations for the conservation, protection and rejuvenation of the Western

Ghats Region following a comprehensive consultation process involving people and Governments of all the concerned States'. The final report of the WGEEP was submitted in August 2011 which is kept secret by the government. On the other hand, the MoEF was preparing additional information requested by the World Heritage Committee with the unpublished WGEEP report.

With regard to the World Heritage nomination, the WGEEP concluded that there was 'a need for greater participation of local people and communities in formulation and implementation of the Western Ghats National Heritage proposal', adding that the 'objections raised at the UN Permanent Forum on Indigenous Issues to the Indian proposals on May 17, 2011' were 'serious and quite genuine'. The Panel also noted that it was 'inappropriate to depend exclusively on Government agencies for constitution and management of ecologically sensitive zones'.

The Panel suggested that instead 'the final demarcation of the Zones (also in context of the UNESCO Heritage Site proposal), and fine tuning of the regulatory as well as promotional regimes, must be based on extensive inputs from local communities and local bodies' and that the 'process of fine tuning the limits of the various zones, deciding on management regimes and the implementation be a participatory process going right down to gram sabhas'. Such an approach, the WGEEP remarked, 'would more effectively serve the objectives of the UNESCO Heritage Programme, than the proposals currently submitted by the Government of India.

Despite these recommendations of the WGEEP, the Government of India went ahead and submitted the Western Ghats World Heritage nomination with clarifications to UNESCO. The additional information submitted by India continued to ignore the Forest Rights Act and the statutory authority of the gram sabhas, and the Government did not 'strengthen community membership and input' as requested by the World Heritage Committee, nor did it 'facilitate increased engagement with all stakeholders to build awareness and support, foster participatory governance approaches, and ensure equitable sharing of benefits'. In response to the Committee's request that India reviews the scope and composition of the serial nomination taking into account the recommendations of the WGEEP, the government claimed that: 'The matter of determining the inclusion or exclusion of sites in the serial nomination has not been dealt by the Western Ghats Ecology Expert Panel and accordingly there are no recommendations on this issue'. In fact, however, the report of the WGEEP did deal with the subject, as outlined above (Bijoy, 2014:233-241).

However, the submitted nomination was totally unfit as it had no prior informed consent of the inhabitants or stakeholders or settlers. In essence, being a member of the Committee, India lobbied hard to make sure its nomination was approved and the Committee meeting was just a formality, a farce devoid of facts or science. The Statement of Outstanding Universal Value (OUV) adopted by the World Heritage Committee states that all component parts of the serial site are 'owned by the State and are subject to stringent protection under laws including the Wildlife (Protection) Act of 1972, the Indian Forest Act of 1927, and the Forest Conservation Act of 1980' (Bijoy, 2014:233-241). There is a fear, which was expressed by local people during the IUCN evaluation and afterwards that the inscription of the Western Ghats on the World Heritage List will have adverse impact on settlers which would precipitate a survival crisis.

The added prestige brought to the site by international recognition and the pressure that inscription entails are likely to perpetuate injustices. Any new management systems for these sites should incorporate the relevant gram sabhas as the authority with power to protect wildlife, forest and biodiversity from 'destructive practices affecting their cultural and natural heritage' in the customary and traditional boundary of *adivasi* villages recognized as a 'community forest resource'. As part of this, the free, prior and informed consent of the relevant gram sabhas must be obtained before any new governance or management mechanisms are introduced, not just as a matter of principle but as an implicit legal requirement under the Forest Rights Act (Bijoy, 2014:241-242).

The Western Ghats therefore contains a significant array of cultural diversity and a diversity of relationships between the inhabitants and the lands on which they depend. Altering the protected status of the Western Ghats would impact on each of these people in a myriad of ways; however, this was not considered in the World Heritage nomination process (Bijoy, 2014:225). The disclosure of WGEEP report has resulted in turmoil across the six states which resulted in constitution of High Level Working Group (HLWG) chaired by Dr. K. Kasturi Rangan. HLWG was appointed to reconsider the recommendations of WGEEP and submitted its recommendations to notify 4,156 villages across WG as ecologically sensitive areas (ESA) (HLWG, 2013:2-3).

ESA is often considered synonymous to: environmentally sensitive areas, environmentally sensitive zones, ecologically sensitive ecosystem, ecologically sensitive sites depending upon the context and the area or location of conservation interest (Gadgil et al., 2011:176). While the constitution of such ESZ/ESAs has had many positive consequences, there are also serious flaws in the system. The most serious problem is that the system depends heavily on bureaucratic regulation. With little or no meaningful participation by the local community and given the absence of bureaucratic transparency and lack of accountability, this breeds corruption. The result is that the weaker sections suffer harassment and extortion, while the wealthy and the powerful successfully flout the regulations, leading to tremendous local resentment (WGEEP, Part I, 2011:27).

It should be noted that the HLWG has categorized the WG into cultural and natural landscapes. The places categorized as natural landscapes, is inhabited heavily. It is not a wilderness area, but the habitat of its people, who share the landscape with biological diversity. It is not possible to plan for Western

Ghats, only as a fenced-in zone, with no human influence. This is the difference between the natural landscapes of a highly populated country like India, against the wilderness zones of many other countries (HLWG, 2013:99).

The environment cannot be protected without the involvement and participation of the local people. Any unreasonable restriction or ban on the rights of the people will have a counterproductive effect. It is an international agenda to expand the protected areas in the developing countries like India for taking the advantage of the Kyoto Protocol which permits carbon trading. Eyeing on foreign funds and auxiliary benefits, NGOs and some officials are promoting environmental colonization with a motive to render farmers and indigenous people as conservation refugees.

The expansion of the buffer areas of the Hampi World Heritage Site and its impact on the lives of the people is a classical example as to how the rights of the people are undermined for meeting international commitments (Campbell, 2015). Having found that the World Heritage Tag could have become an impediment to the development plans, the Government of India on May 21, 2015 had withdrawn the Delhi City nomination for the UNESCO World Heritage City Tag (Verma, 2015). This is equally applicable to the sites across Western Ghats inscribed on the list of World Natural Heritage Sites as it could affect the entire development as well as growth of the region and the survival of the inhabitants.

The problem which could be faced by the people in an ESA can be read out from the Mahabaleswar-Panchagani experience in Maharashtra. The Mahabaleswar-Panchagani ESA was notified in 2011 and the same has become an impediment for all the development activities (Kapoor et al., 2009). The lives of the people living there are virtually at the mercy of various government officials and NGOs. We cannot undermine the real human problems relating to the conservation initiatives by notifying an area as ESA.

Conclusion

The scenario of global warming and immediate threats of climate change requires proper governance and conservation of environment. The national and regional environmental governance mechanisms are inextricably connected with the international regimes of conservation due to the trends of globalization. When the international regimes influence the national governance and conservation policies, there occur the differences between conservationists and the inhabitants of the concerned region. Most of the time, environmental governance and conservation are faced with the challenges of determining how to shape human interactions with nature in landscapes of which people are a part and the fight for the existence of the stakeholders.

The contribution of Western Ghats ecosystem to the mitigation of climate change is extremely prominent and attracted the attention of international climate change funding. But the success as well as contradictions of the environmental governance in trapping the human capital which is huge and extraordinarily well equipped in terms of education and motivation in the Western Ghats region to undertake conservation action is in question with the Gadgil and Kastoori Rangan reports. The conflict arises when the strategic directions seek to catalyze change and undertake innovative and effective approaches to environmental governance without the involvement of tremendous social and human resources of the region.

In reality, environmental governance and biodiversity conservation as ecologically sensitive areas (ESAs) in India are zones of bureaucratic activism—instead of taking into account public faith and make sure of their participation in governance and conservation. Western Ghats is a thickly populated region which needs a governance system and conservation agenda of inclusive growth of people as well as environment. This requires conservation modalities and procedures with effective and practical tools promoting a rights-based approach such as disclosure of environmental assessments and channels for public participation. In each and every instance, the success of long-term sustainable governance of natural resources is contingent upon local people's support. Therefore, considering local people's attitudes, taking into account their requirements, and honouring their opinions, should become a priority in environmental governance and conservation.

References

Anonymous (2010), "Clarifications Sought on Western Ghats", *The Hindu*, Monday, October 18.

Bawa, Kamal S., Arundhati Das, Jagdish Krishnaswamy, K. Ullas Karanth, N. Samba Kumar and Madhu Rao (2007), "Western Ghats and Sri Lanka Biodiversity Hotspot: Western Ghats Region, Ecosystem Profile, Critical Ecosystem Partnership Fund", Conservation International, USA.

Bijoy, C.R. (2014), "Western Ghats of India: A Natural Heritage Enclosure?" in Stefan Disko and Helen Tugendhat (eds.), 'World Heritage Sites and Indigenous Peoples' Rights', Denmark: IWGIA, Forest Peoples Programme, Gundjeihmi Aboriginal Corporation.

Brockington, Dan, Jim Igoe and Kai Schmidt-Soltau (2006), "Conservation, Human Rights, and Poverty Reduction", *Conservation Biology*, 20(1): 250-252.

Campbell, Morgan (2015), "Hampi Bazaar Demolition II: How Maps Alienate People", *Economic and Political Weekly*, L (29), available at: http://www.epw.in/reportsstates/hampibazaardemolitionii.html.

CIEL (2011), "Climate Change and Human Rights: A Primer", The Centre for International Environmental Law, Geneva.

Equations (2011), "Tourism in Forest Areas of Western Ghats", paper submitted to the Western Ghats Ecology Expert Panel (WGEEP), Bengaluru.

Gadgil, Madhav, R.J. Ranjit Daniels, K.N. Ganeshaiah, S. Narendra Prasad, M.S.R. Murthy, C.S. Jha, B.R. Ramesh and K.A. Subramanian (2011), "Mapping Ecologically Sensitive, Significant and Salient Areas of Western Ghats: Proposed Protocols And Methodology", *Current Science*, 100(2): 175-182.

HLWG (2013), "Report of the High Level Working Group on Western Ghats: Volume I", Ministry of Environment and Forests, Government of India.

Ivanova, Maria (2007), "International Environmental Governance Reform: Options and Implications", Draft Report, International Environmental Governance, Chatham House Workshop, London.

Kanagavel, Arun, Shijo Joseph, Revati Pandya and Rajeev Raghavan (2013), "Potential for Community and Conservation Reserves in the Western Ghats of India", *Asian Journal of Conservation Biology*, 2(1): 61-68.

Kapoor, M., K. Kohli and M. Menon (2009), "India's Notified Ecologically Sensitive Areas (ESAs): The Story So Far", *Kalpavriksh*, Delhi and WWF-India, New Delhi.

Kothari, Ashish and Neema Pathak (2010), "Indigenous and Local Community Based Conservation in India: Current Status and Future Prospects", in Kristen Walker Painemilla, Anthony B. Rylands, Alisa Woofter and Cassie Hughes (eds.), 'Indigenous Peoples and Conservation: From Rights to Resource Management', Arlington, VA: Conservation International.

Lawson, Andrew and Neil Bird (2008), "Government Institutions, Public Expenditure and the Role of Development Partners: Meeting the New Challenges of the Environmental Sector", Final synthesis report, ODI, London, March.

OECD (1999), "Donor Support for Institutional Capacity Development in Environment: Lessons Learned", OECD Publishing, Paris.

OECD (2012), "Greening Development: Enhancing Capacity for Environmental Management and Governance", OECD Publishing, Paris.

Sharma, Anju (2009), "Planning to Deliver: Making the Rio Conventions More Effective on the Ground", Climate Change, Biodiversity, Desertification, GTZ.

Singh, Asha (2014), "Environmental Governance in Small Island Developing States", United Nations Environment Programme (UNEP), Nairobi, Kenya.

Slunge, Daniel and Emelie César (2010), "Swedish Bilateral Support to Environmental Capacity Development: Overview of Key Results and Lessons Learned", Environmental Economics Unit, Department of Economics, University of Gothenburg.

Thornton, Nigel (2010), "Making the Most of Climate Change Finance in Asia and the Pacific", A synthesis report from five country studies in Bangladesh, Cambodia, Indonesia, Philippines and Vietnam, facilitated by the Capacity Development for Development Effectiveness Facility, CDDE.

UNEP (2005), "Register of International Treaties and Other Agreements in the Field of the Environment", UNEP/Env.Law/2005/3; Division of Policy Development and Law Environmental Law Branch United Nations Environment Programme.

UNFCC (2015), "Adoption of a Protocol, Another Legal Instrument, or An Agreed Outcome with Legal Force under the Convention Applicable to all Parties", Conference of the Parties, Twenty-first Session, Paris, November 30 to December 11, 2015; Agenda item 4(b) Durban Platform for Enhanced Action (decision 1/CP.17).

Verma, Richi (2015), "Heritage Tag Could Have Slowed Growth", *Times of India*, available at: http://timesofindia.indiatimes.com/city/delhi/Heritage-tag-could-have-slowed-growth/articleshow/47392252.cms.

WGEEP Part I (2011), "Report of the Western Ghats Ecology Expert Panel Part I", Submitted to The Ministry of Environment and Forests, Government of India.

Wingqvist, Gunilla Olund, Olof Drakenberg, Daniel Slunge, Martin Sjostedt and Anders Ekbom (2012), "The Role of Governance for Improved Environmental Outcomes: Perspectives for Developing Countries and Countries in Transition", Sweden: The Swedish Environmental Protection Agency.

8

Plight of the Adivasis in Kerala

K.M. Seethi

Muthanga was a landmark in the adivasi (tribal) land struggle in Kerala. 'Occupy Forest' Movement (OFM) launched by the Adivasi Gothra Mahasabha (AGMS) in early 2003 turned out to be a catalyst for political mobilisation of the marginalised communities in the state (Seethi, 2003). 14 years have gone by since the OFM became a lesson for introspection and agenda setting in respect of the adivasis of Kerala. The OFM experience also calls into question the complex dimensions of 'identity politics' and the mobilisation of the caste/tribal groups in a state which is known for public action and civil society intervention on questions of social oppression and exclusion. The paper tries to revisit the OFM and the continuing plight of the adivasis in the post-Muthanga period.

Identity politics seeks to sustain the politics of rights in which 'identity' emerges as a mobilisational instrument for achieving internal cohesion and solidarity with respect to the groups in action (Kumar and Seethi, 2014-15). As such, it is basically a politics of resistance and mobilisation. In India, the most visible expression of identity politics could be seen among the movements and struggles of the lower castes and tribes, who were historically marginalised from the mainstream society (Beteille, 1986:297-318). The adivasi community (the scheduled tribes) here remain at the bottom of the social hierarchy whose lives have been worsened by the increasing incidence of poverty, illiteracy, ill-health, unemployment, powerlessness and other forms of human deprivations. The scheduled castes and scheduled tribes in India face marginalisation and social exclusion in a wide variety of forms due to development, displacement and discrimination (Burman, 1992; Dubey and Ratna, 1977).

The factors that have significantly affected the life-world of the adivasis are their relations with land and forest. The history of the encroachment on adivasi land goes back to the British period. With the introduction of private property in land and the emergence of the social forces in diverse areas of the society, large scale alienation of land took place (Kunhaman, 1989; Kunhaman, 1985). Though measures were initiated, from time to time, to put an end to land encroachment in the post-Independence period, the process continued unabated and even worsened. The most important cause of adivasi land alienation was the penetration of non-adivasis into the adivasi land, besides the developmental activities started by the state.

The adivasis of Kerala, over years, were forced to organise themselves though they constituted only a small minority of 1.10 percent of the State's population (India, MHRD, 2001; Luiz, 1962). As a result of land encroachment, more than three-fourth of the adivasis became landless and their social conditions were far below the state average (Seethi, 2003; Bijoy, 2003a; Bijoy, 2003b; Kjosavik, 2006). Hence, the adivasi question in Kerala needs to be understood within a larger socio-economic setting and the structural conditions prevailing in the state. The OFM in Muthanga also has history of betrayals by both the state and the settlers in the adivasi lands.

Muthanga Struggle

For the 3.64 lakh indigenous adivasis of Kerala, what happened on February 19, 2003 in Muthanga, the Wayanad district of Kerala, was nothing short of state violence in its worst form. With the violence being unleashed on a large scale, the adivasis became 'humans without human dignity' and without the basic right to life. Muthanga had even drawn the attention of the world on account of the violence perpetrated on the unarmed indigenous people—that too when only months were away for the finale of the International Decade of the World's Indigenous People (1995-2004). For the adivasis of Kerala, the decade that was just passing over was a decade of empty promises and betrayals, and hopes and frustrations culminating in a witch-hunt assuming violent proportions (Seethi, 2003; Bijoy, 2003a; Bijoy, 2003b).

This was the first time in the history of Kerala that the adivasis, who continued to get driven from pillar to post, became the target of police firing and brutality which resulted in deaths and destruction. The casualties of the police firing were, at that point, shrouded in mystery. Questions were also raised as to why the media personnel were not allowed to enter the area when the operation was underway against the adivasis. During the operation, which lasted for almost 18 hours, the police cordoned off the entire area creating a panicky situation. Many adivasis and human rights activists believed that the police as they went berserk within the forest area would have buried all evidences of its brutality (Seethi, 2003; Bijoy, 2003a; Bijoy, 2003b).

The Muthanga operation came as a sequel to the plans chalked out by the forest officials and police, supported ostensibly by the settler-politician-forest mafia nexus, which was determined to drive the adivasis out to hold their sway in the area. Hundreds of adivasis belonging to the AGMS led by C.K. Janu and M. Geethanandan had occupied the forestland in Muthanga, which was under the Wayanad Wild Life sanctuary (WWLS), in early January 2003 in protest against the government inaction on the promises made to them in 2001. The AGMS had made it clear from the beginning that the tribals would leave the forest only after the government conceded their demands. The leaders also insisted that any further discussion on the land issue could be held only with the Chief Minister.

The adivasis had put up huts and other shelters in the area and announced that they would start cultivating the occupied area. However, even after weeks of their occupation of the forest land, the government chose to remain silent. In fact, the adivasi leaders were expecting a call from the State capital to start talks on the land issue. But the State cabinet, which met a few days before the Muthanga operation, had decided "not to take any immediate steps to settle the issue" as "the time was not ripe for talks". This was seen as a criminal negligence of a democratically elected government (PUCL, 2003; Prabhakaran, 2003).

Even as the agitation was making headway, a section of environmentalists and activists demanded the immediate eviction of the 'encroachers' from the protected forest, as the areas the adivasis had occupied formed part of the Nilgiri biosphere of the Western Ghats, one of the 25 biodiversity hotspots of the world. The area was also adjacent to the wildlife sanctuaries of the Bandipur Tiger Reserve and Project Elephant Area and the Mudumalai Reserve Forest. However, they did not care to see how the land had already become barren in the areas occupied by the adivasis. This was exactly the area where the forest mafia and settlers had inflicted severe damage to the sanctuary. More importantly, the occupied part of the forest was already in a degraded condition because the State government had let the plantation of eucalyptus trees in hundreds of acres for the benefit of the Birla-owned Grasim Industries, which had already left Kerala after causing tremendous damage to the State's environment (Seethi, 2001; Seethi, 2000).

The plantation of the eucalyptus had resulted in large-scale ecological destruction and drying up of streams in the forest. For more than a decade, the wild animals had not been coming to this area for sustenance due to shortage of water and food. The adivasis accused the State Forest Department of turning this sanctuary into a wasteland and drying up of water bodies. They, however, assured that the tribals would only initiate plans for the ecological restoration of the degraded forest in the sanctuary. The environmentalists just pooh-poohed such assurances of the adivasis. And they were also not worried about the question where all the money received by the Forest Department from the World Bank for greening the sanctuary had gone.

Instead of engaging the adivasis in a democratic manner, the authorities resorted to all sorts of intimidation, and unleashed a vilification campaign against the leadership and the struggling people. Sensing this, the AGMS leaders had warned, almost a month before the operation, that there was a conspiracy to sabotage the agitation by causing damage to the sanctuary occupied by them, thereby shifting the entire burden of the ecological destruction on the shoulders of the adivasis, which could also be used as a pretext to evict them (Janu and Geethanandan, 2003). Their fears came true. This was exactly what happened in a few weeks time in and around Muthanga. By the middle of February 2003, a fire broke out inside the sanctuary under mysterious circumstances.

The adivasis believed that this was deliberately done by some people, with the connivance of the forest officials, to smoke them out. Naturally, the disgruntled adivasis took captive of some forest officials who had gone to the sanctuary to 'inquire' into the forest fire. However, these 'officials' were released next day on condition that the District Collector would personally record their statement on the circumstances under which they were taken by the Adivasis (Seethi, 2003). The AGMS alleged that the people held hostage by them were engaged by the Forest Department to set fire in the sanctuary in order to tarnish the image of adivasis, thereby creating a feeling that their presence in the forest would only cause damage to the rich biodiversity of the sanctuary and wildlife (Janu and Geethanandan, 2003).

Tension mounted at Thakarapadi where the adivasis had put up a check post to restrict the entry of the outsiders. The local people assembled in the area swelled their ranks and turned hostile demanding immediate action against the AGMS leadership. Capitalising the aggressive mood of the local people and the agitating adivasis, the police launched its operation on February 19 under instructions from the higher-ups. For more than 1,000 adivasis living in the area for a month and a half, it was a day of horror and unanticipated feral mêlée. They never expected such an organised violence of the State—that too on a sensitive issue about which the Chief Minister himself had shown some 'sympathy' in the past.

The operation started in the morning of February 19 with a large contingent of police and Forest Department personnel entering into the sanctuary and demolishing all that they had seen on the way. The check post, thatched huts, makeshift tents, household utensils etc. were destroyed and set afire. The police did not spare even women, children and aging adivasis who were running away from the scene of terror. Their kith and kin got scattered around the forest and the police chased them away to interiors. Cruelty knew no bounds. Even those who were unable to move after the brutality received severe injuries because of repeated beating with rifle butts (Seethi, 2003; Janu and Geethanandan, 2003; Bijoy, 2003a; Bijoy, 2003b).

Elsewhere in the sanctuary, pitched battles broke out when some tribals came face to face with the aggressive forces. When a section of the struggling adivasis refused to disperse even after teargas shells were fired, the police resorted to indiscriminate firing of several rounds. Meanwhile, one policeman who was taken hostage by the tribals died of injuries. Reports indicated that the adivasis had pleaded for medical facilities for the injured, including their own men, but the police turned it down as if they were determined to sacrifice one of their own men to rationalise its offensive action. At this time, the police confirmed the death of one tribal activist in the firing and the policeman who sustained injuries under captivity.

But the adivasi leaders feared that there would have been more causalities which could not be confirmed due to the total blackout declared by the police, which lasted for almost 18 hours. Many suspected that during those critical hours the bodies of those killed would have been buried in the forest and all evidences might have been destroyed. A probing mission led by Arundhati Roy had long conversations with the victims of the police atrocities. They were reported to be under the trauma of the police action. The adivasis who were admitted to various hospitals later told that they had apprehensions about their kith and kin. Those who had been taken to custody after the operation also said that their children and relatives were missing. Many adivasis who had joined the struggle from other parts of the State had not returned to their places. All this strengthened the fear that there would have been more casualties (Seethi, 2003).

The most unfortunate turn of the struggle was that the State government all of a sudden emerged hostile towards the adivasis, and the Chief Minister himself defended the police terror saying that it came as a "last resort when the adivasis went on to attack police and forest officials". The government failed in its primary responsibility to get accurate information on the ground situation in those crucial hours, by leaving everything in the hands of corrupt officials and rude police force to handle things in their own ways. Instead of eliciting information from them, the Chief Minister should have visited the area and verified things in person. In his first statement on the Muthanga operation, the Chief Minister made some comments on the struggling adivasis saying that there were "ulterior forces" behind "the tribal attacks".

To everyone's embarrassment, even the left and radical forces in the state alleged that there were "forces behind" the agitation inimical to the "national interests". This was a regular pattern of accusations that one was familiar with, particularly when democratic struggles would have no direct patronage of any mainstream political party. Any spontaneity in action would even mean "behind-the-scene sponsorship" of naxalites, People's War Group, LTTE etc. The superciliousness of the State found its full expression when the Speaker of the State Legislative Assembly said that such struggles should be dealt with force and bullets. This was what a constitutionally elected person occupying the august office of the State told the legislature in response to an adjournment motion raised by the opposition. The intervention of the National Human Rights Commission, the National Commission for Scheduled Caste and Scheduled Tribe and many other human rights organisations had an impact on the position of the government for instituting an impartial inquiry (Seethi, 2003).

Many rights activists pointed out that the government of Kerala shirked its constitutional and statutory obligations. The rights of adivasis are granted protection under the Protection of Civil Rights Act, 1955 and the Scheduled Castes and Scheduled Tribes (Prevention of Atrocities) Act, 1989. The 1989 Act provides for punishment for the destruction of property belonging to scheduled caste or scheduled tribe communities as well as for legal action against public servants who wilfully neglect their duties required to be performed under the Act. It also provides for special courts to be set up to hear cases of violations. That the government's neglect in implementing the provisions of this Act had led to a widespread feeling of impunity amongst those committing atrocities against these indigenous communities. Reports indicated that the majority of crimes committed against members of these communities were charged under sections of the IPC rather than the 1989 Act, which would provide better protection for these communities and harsher punishments for the violation of their rights (Seethi, 2003; Janu and Geethanandan, 2003; Bijoy, 2003a; Bijoy, 2003b).

Background of the OFM

The circumstances that led to the police firing and subsequent brutal 'follow-up' operation need to be looked into against the background of the developments that commenced since 1975 when the State Assembly adopted legislation on land alienation of adivasis (Kerala Law Department, 1975). It is true that the adivasis of Kerala had become more articulate and assertive about their rights. They also became aware of the laws and the ways of the non-tribal people, politicians, governments and the courts in addressing their basic question. One must bear in mind that the adivasis had at their disposal large tracts of forests in the State, particularly in Palakkad, Wayanad, Idukki, Pathanamthitta, Kollam and Thiruvananthapuram districts. These lands were slowly but periodically taken over by non-adivasi people, and successive governments were only passive spectators while the encroachment drive was steadily underway. As the pressure on land had increased in the plains, non-adivasi settlers further went ahead with land grab (Seethi, 2003).

The adivasis were thus perpetually cheated by the settlers. Their ignorance and innocence were capitalised by settler farmers. The latter used both threats and persuasions on the adivasis to relinquish their ancestral land for whatever they offered. In a majority of cases, documents validating such transfers were not available. And there were instances when the adivasis had to sign on blank sheets of paper provided by the settlers which were later used to manipulate ownership. Thus, the non-adivasis who secured possession of the lands eventually became the 'owners'. Having been kicked out from their ancestral land, the adivasis soon became languished in grinding poverty which compelled them to look for other areas for food and shelter. As they sought refuge in the new stretches of forest land, the non-tribal settlers again emerged to disrupt their life. In the midst of this chase, they were in fact searching for a civil society to provide them a leg up. The adivasis knew very well that all political parties, irrespective of political affiliations, would only ignore their plight, giving a silent approval of the appropriation of the adivasi land insofar as the settlers had become a major vote bank (Seethi, 2003; Janu and Geethanandan, 2003; Bijoy, 2003a; Bijoy, 2003b).

As stated earlier, the year 1975 was a watershed for the marginalised adivasis of the State. In April 1975, the State Assembly unanimously passed the Kerala Scheduled Tribes (Restriction on Transfer of Lands and Restoration of Alienated Lands) Bill, which was to prevent the lands of the Adivasis from falling into the hands of non-tribals. The 1975 Act had assured the adivasis that their alienated land would be given back and that all transactions of adivasis land after January 1, 1960 would become null and void (Kerala Law Department, 1975). The 1975 Act got the Presidential assent in November that year and was incorporated in the Ninth Schedule of the Constitution, which ensured that the Act could not be challenged in any court of law. It was apparently a realisation of a long-cherished dream of the adivasis. However, the euphoria was short-lived. All governments that came to power in Kerala since 1975 were still reluctant to implement the provisions of the Act, obviously under pressure from the settlers.

Meanwhile, the State government formulated rules for restoration of alienated lands in 1986, but it specified that the 1975 Act would come into effect retrospectively only from January 1, 1982. The Act stipulated that the adivasis should return the amount, if any, they had obtained during the original transaction and pay compensation for any improvements made on the land by the non-tribal occupants. This amount was to be advanced by the government to the tribal people as loans and recover from them in 20 years. As per this, only 8,500 applications were filed by the adivasis, as a good majority of them were not acquainted with the new legislation or were reluctant to receive loans or were bribed by the corrupt settler-bureaucratic nexus. Consequently, the scenario of exploitation continued and it only helped encourage the settlers to carry on their occupation of the adivasi land. It was in such a circumstance that Nalla Thampi Thera, a non-adivasi from Wayanad district, filed a case in the High Court of Kerala seeking a direction to the State government to implement the 1975 Act. The High Court gave its verdict in 1993 asking the government to implement the Act within six months. However, the government again went on delaying the process for two and a half years seeking extensions of deadline to implement the Act (Seethi, 2003; Janu and Geethanandan,

2003; Bijoy, 2003a; Bijoy, 2003b).

In 1996, the High Court set a final deadline of September 30, 1996 to evict the non-adivasi encroachers, if necessary seeking the help of the police, and reminded the officials concerned that contempt of court proceedings would be initiated if they failed to implement the court directive. Surprisingly, the government resorted to a backdoor method and came up with another controversial bill diluting the original provisions of the 1975 Act, ostensibly to help the settlers. This was the time when the adivasis became increasingly disillusioned with the promises of the government. Notwithstanding this, the State Assembly passed the Kerala Scheduled Tribes (Restriction on Transfer of Land and Restoration of Alienated Lands) Amendment Bill, 1996 almost unanimously with just one dissenting vote. The Bill sought to legitimise all transactions of tribal land up to January 24, 1986, the implication of which was that the government was not obliged to restore all alienated land as per the 1975 Act. The government tried to defend the new Act by arguing that it was the "only practical alternative" (Bijoy, 1999).

However, the President of India did not give his assent to the 1996 Amendment Bill on the grounds that the 1975 Act had already been incorporated in the Ninth Schedule of the Constitution. To sidestep this, the State Assembly passed another Bill, The Kerala Restriction on Transfer by and Restoration of Lands to Scheduled Tribes Bill, 1999 (Kerala Law Department 1999). It defined 'land' as 'agricultural land', a State subject, in order to avoid sending it for presidential assent. The new Bill also sought to repeal the key provisions of the 1975 Act. For instance, the amended Act stipulated that only alienated land in excess of two hectares possessed by encroachers would be restored, while alternative land, in lieu of the alienated land not exceeding two hectares, would be given elsewhere. An impression was that the number of applicants claiming land in excess of two hectares would be much less, making restoration unnecessary. The 1999 Act also had a provision to provide land to other landless tribal people within two years. Apparently, there were about 11,000 such families in the State. Meanwhile, the High Court of Kerala rejected both the 1996 and 1999 Acts and declared the provisions under them illegal. The State government went on appeal to the Supreme Court and secured stay orders. This was the background of the adivasi mobilisation (Seethi, 2003; Janu and Geethanandan, 2003; Bijoy, 2003a; Bijoy, 2003b).

It was estimated that there were 3.64 lakh adivasi people in Kerala (or about 70,000 families) (Kerala, Directorate of Census Operation, 2001). Evidently, as many as 45,000 families were landless. Under the provisions of the 1999 Act, there were only 4,500 applications, which meant a good majority of the landless adivasis would not come under the purview of the new Act. In fact, the starvation deaths reported during this time were from these landless adivasis. For example, when the adivasis launched their agitation in August 2001, not less than 32 adivasis had died of starvation in a few weeks time. The agitation was in the midst of that year's Onam festival, when starvation deaths were reported in the State. Led by the Adivasi-Dalit Agitation Committee, hundreds of adivasis started streaming into the State capital from the forests and mountain regions and had put up huts in front of the Chief Minister's official house as well as the State Secretariat. Their major demand was that 45,000 landless adivasi families be given five acres each of cultivable land. However, it took more than a month and a half for the state government to take a decision on the land issue (Bijoy, 2003a; Bijoy, 2003b).

The adivasi struggle had then attracted considerable public attention. This was for the first time that an organised agitation of the adivasis had taken place anywhere in the State. More importantly, the struggle was led by a tribal woman, C.K. Janu, who, more than anybody else, had brought the adivasi rights to public attention, as also the land grab and sexual exploitation of their women (Janu, 2004). Even as the agitation was underway, the government had agreed to undertake development programmes for adivasis. However, the land issue was not adequately clarified. The government initially offered 10,000 acres in two months. But the government did not specify about the locations, the type of land or the number of beneficiaries. However, the adivasis insisted on "land for livelihood". After sometime, the State government announced that it had found 5,000 acres more for distribution, but eventually said that only 10,000 acres would be distributed in accordance with the provisions of the 1999 Act. Adivasis considered it as an illegal Act intended to deny their rightful land. They refused to buy the government argument that sufficient land would not be available for distribution. They said that the government had been illegally holding on to all land that was sanctioned for the adivasis on many occasions. Hence, their major demand that all landless adivasis must get five acres each (Seethi, 2003; Bijoy, 2003a; Bijoy, 2003b).

At this stage, the adivasi leadership had come to the realisation that their attempts to recover alienated land from the settler farmers would result in a new set of contradictions given the fact that all political parties had a common stake in the 'affairs' of the settler-farmers. Adivasi leadership itself pointed out that all governments were playing hide and seek on this question by pointing to the issue of 'tension' between adivasis and settler-farmers. Meanwhile, they were reported to have identified 11 lakh acres of land that could be easily available for distribution. It may be recalled that the government needed to distribute only less than 3 lakh acres to provide land to all landless adivasis in the State (Seethi, 2003).

Pointing to the bitter experiences in the past, many had argued that the adivasi demand "land for all" might result in large-scale illegal transfer of adivasi land to non-adivasis. In response to such positions, the adivasi leadership maintained that it was a typical argument for further procrastination of the basic issue and that the government could easily ensure it would not happen, by making another law to protect that land. In fact, the 1975 Act itself had provisions to check any further alienation. Evidently, the land adivasis had been demanding all these years was absolutely free of any legal implications. While the successive governments had been so generous to settler farmers and forest-politician mafia, they became so penny-pinching when adivasis demanded five acres. This, in fact, had provoked the struggling adivasis.

It was at this time that the adivasis decided to form the Adivasi Dalit Samara Samithy to spearhead the movement. The Samithy set up a council of leaders representing the 34 tribal communities in Kerala that eventually led to the formation of the Adivasi Gothra Mahasabha (Adivasi parliament) on October 3, 2001 at Thiruvananthapuram. The AGMS sought to mobilise support not only from the adivasi communities but also from broad sections of the civil society. The struggle went on for more than a month and a half (Raman, 2002). But the government did not show any interest in addressing the issues raised by them. This was the background of C.K. Janu's declaration of a fast unto-death. Facing widespread public criticism, the government then decided to initiate talks with the leaders.

The 48-day old struggle of the tribal people in Kerala ended on October 16, 2001 in the wake of a seven-point agreement between the State government and the Adivasi Dalit Action Council. The agreement states:

1. Wherever possible, the government would provide five acres of land to each landless adivasi family; at other places, the offer is a minimum of one acre, which may go up to five acres depending on the availability of land;
2. A five-year livelihood programme would be implemented in the land thus provided until it becomes fully productive for adivasis to sustain themselves;
3. The State would enact a law to ensure that the land provided to adivasis is not alienated as had happened in the past;
4. The government would soon pass a resolution asking the Union Government to declare the adivasi areas in the State

as scheduled areas, bringing them under Schedule V of the Constitution;

5. The government would abide by whatever decision the Supreme Court takes on its appeal against the Kerala High Court order quashing the unpopular law (the Kerala Restriction on Transfer by and Restoration of Lands to Scheduled Tribes Bill, 1999) passed by the State Assembly in 1999;
6. The government is to implement a master plan for tribal development and the plan is to be prepared with the participation of adivasis; and
7. The maximum possible extent of land will be found and distributed in Wayanad district—at least 10,000 acres—where there is the largest concentration of landless Adivasis (Seethi, 2003; Bijoy, 2003a).

Many called the conclusion of the agreement as a 'historic triumph' and 'a morale booster' for similar struggles across the country. It is true that the agitators' main demand that all landless adivasi families be provided with five acres each was not fully conceded. Yet, the agitation was seen as a near success given the fact that for the first time landless adivasis in Kerala got a positive commitment from the government on, at least, one acre of land. They were also to get the protection of a new law preventing any further alienation of their land. The most significant aspect of the agitation was that the agenda of the five decades-old adivasi struggle in Kerala had been changed from the "restoration of alienated land" to "land for the landless tribal people" (Seethi, 2003).

State Chief Minister A.K. Antony announced, after the agreement with the adivasis, that the government would try to get 10,000 acres of land in Wayanad where the number of landless adivasis was the highest, besides the 42,000 acres of land already identified for this purpose in different parts of the State. The Chief Minister said the government would stand by its earlier offer to provide 90 percent of employment in forests to the adivasis. C.K. Janu said the Adivasi-Dalit Action

Council was happy that the government had accepted almost all its demands. However, the Council, she said, was aware of the difficulties in finding sufficient land instantly to provide five acres to each landless adivasi family. So, it had decided to accept the government offer to provide five acres where land was available and between one acre and five acres in other places (Bijoy, 2003).

For the fourteen months since October 2001, till the AGMS occupied lands in Muthanga, the government went on delaying the implementation of the agreement. Though the government had officially started land distribution, it turned out to be a farce. By April 2002, 568 families were provided with 1,308 acres of land when the Chief Minister, along with C.K Janu of AGMS, began the first land distribution at Marayur in Idukki District. That means just 1.06 percent of the families were provided 2.2 percent of the identified land within the first four months of the period earmarked to complete the task. However, nothing much had happened after that, except the report appeared in the media that the land actually allotted was 1,770 to 848 families, i.e., only 3 percent of the promised land. The government continued its negligence by bringing in numerous legal and administrative hurdles such as the dispute between the Forest Department and Revenue Department and the problems associated with reserve forest areas and the Forest Conservation Act, 1980 (Bijoy 2003a; Bijou, 2003b).

It is true that the Forest Conservation Act, 1980 prohibits any encroachment into the forests since 1980. When forests became part of the Concurrent List, the State government could denotify forest lands only with the prior consent of the Union Government. However, the 1980 Act prohibited encroachments after 1980 only. The implication was that the encroachments before 1980 should not be treated as 'encroachments' but need only be resolved as per the law. Those who had claims or rights to the forest areas prior to 1980 should have been provided that as per law (Bijoy, 1999).

The State government obviously did not have the political will to fulfil its constitutional and statutory obligations in regard to the adivasi land. The State government had wriggled out its constitutional responsibility in other important realms too. No adivasi land in Kerala had been declared as scheduled area, thus depriving the adivasis of their entitlement in accordance with the provisions of the Constitution of India, even as other States such as Maharashtra, Gujarat, Rajasthan, Madhya Pradesh, Bihar, Jharkhand, Chattisgarh, Odisha, Himachal Pradesh and Andhra Pradesh had gone ahead with that. The non-inclusion of adivasi land under the scheduled area also delayed the implementation of the provisions of Panchayats (Extension to the Scheduled Areas) Act of 1996 through necessary legislations at the State level. The OFM was the culmination of all these neglect and denials.

As the public opinion snowballed in favour of the adivasis after Muthanga, particularly with the intervention of different agencies such as National Human Rights Commission (NHRC), the government of Kerala had to order an independent inquiry. Thus, the Central Bureau of Investigation (CBI) stepped in for a comprehensive inquiry into the police firing and the circumstances leading to violence. But the adivasis, having been disappointed subsequently, alleged that it was a sham investigation because the CBI report concluded that the adivasis were "trespassers". Later, the CBI even justified the forcible eviction of adivasis, saying that the police had taken action following untoward incidents (Seethi, 2003).

The Muthanga agitation certainly enhanced the political consciousness of the adivasis in the state. After that, even mainstream political parties began capitalising the adivasi sentiments. The Muthanga struggle had also strengthened the Dalit political consciousness in the state. Aralam and Chengara land struggles need to be viewed in this background (Rammohan, 2008; Sreerekha, 2010). C.K. Janu and others argued that as a result of the mobilisation, hundreds of landless families benefited. They said that the dominant parties in the state who wielded political power from time never did nothing to provide land to adivasis. During the post-Muthanga period, however, they were compelled to talk about land rights of adivasis. AGMS believed that they acted as a catalyst for setting an agenda of 'land for landless'.

Conclusion

A decade has gone since the Muthanga struggle, and leaders like C.K. Janu conceded that they could not do much to carry forward the struggle. However, according to them, the struggle could evoke considerable attention and sympathy from the wider public realm of the state that continued to influence the decisions of the government on questions related to dalits and adivasis. The most enduring result of the Muthanga struggle was that the adivasis began asserting as a collective social entity with increasing consciousness about their right to existence in the society. The adivasi leaders, however, admitted later that there were also symptoms of disintegration of adivasi solidarity. This was mainly due to the continuing apathy of the state and their existential struggle for survival and livelihood. The adivasis found themselves difficult to survive if they did not go for their daily work. For some sections, the Muthanga struggle also offered some bitter lessons.

Notwithstanding all these compelling factors, adivasi leaders believed that they became critically conscious of their rights, dignity and distinctiveness. Yet, even after several years of the Muthanga struggle, the restoration of land to the adivasis continued to be a major pending question. Before Muthanga a mere 1.06 percent of the families were provided with land within the first four months of the stipulated 12 months period. There has not been any major change in these figures in spite of the efforts underway. This was obviously one of the reasons as to why a leader like C.K. Janu opted to be a part of the BJP-led National Democratic Alliance (NDA), and even contested with the BJP support.

For the Sangh Parivar, the State Assembly election 2016 was a crucial test of its capability to break through the bipolar coalition dispensation. Hence, it sought to bring in a rainbow coalition of various caste-communal groups in the state. This paid dividend to the third front by way of increasing its percentage of votes as well as opening an account in the State legislature (Seethi, 2016). Admittedly, the entire focus of the adivasi question in Kerala is their right to live with dignity in their lands. The adivasi movement today faces the most crucial existential crisis though no one could ignore the reality of their mobilisation as a strategic form of identity politics. However, the tragedy in Muthanga will go down in history as yet another instance of the self-deception of the successive state governments, political parties and other forces in the state.

References

Beteille, Andre (1986), "The Concept of Tribe with Special Reference to India", *Journal of European Sociology*, 27: 297-318.

Bijoy, C.R. (1999), "Adivasis Betrayed: Adivasi Land Rights in Kerala", *Economic and Political Weekly*, XXXIV (22): May 29-June 4: 1329-1335.

Bijoy, C.R. (2003a), "The Adivasis of India: A History of Discrimination, Conflict, and Resistance", *PUCL Bulletin*, February.

Bijoy, C.R. (2003b), "Adivasis Shot Dead by Police in Wayanad", *PUCL Bulletin*, April.

Burman, Roy B.K. (1992), "Indigenous Peoples and Their Quest for Justice", in B. Chaudhuri (ed.), 'Tribal Transformations in India', Vol. 3, Delhi: Inter-India Publication.

Chandran, V.P. (1999), "Land Settlement and Agrarian Unrest in Wayanad", in E.K.G Nambiar (ed.), 'Agrarian Indian: Problems and Perspectives', Calicut: Association for Peasant Studies, University of Calicut.

Damodaran, A. (2006), "Tribals, Forests and Resource Conflicts in Kerala, India: The Status Quo of Policy Change", *Oxford Development Studies*, 34(3).

Dubey, S.N. and Ratna, Murdia (1977), "Land Alienation and Restoration in Tribal Communities in India", Bombay: Himalaya Publishing House.

Ministry of Human Resource and Development (MHRD) (2001), "Census of India, 2001: The Scheduled Tribes", New Delhi: Government of India.

Ministry of Tribal Affairs (2007), "Annual Report, 2006-07", New Delhi: Government of India.

Ministry of Tribal Affairs (2011), "Annual Report, 2010-11", New Delhi: Government of India.

Indian Institute of Management (2006), "Wayanad Initiative: A Situational Study and Feasibility Report for the Comprehensive Development of Adivasi Communities of Wayanad", submitted to Scheduled Castes and Scheduled Tribe Development Department, Government of Kerala, Centre of Excellence, Indian Institute of Management Kozhikode KIRTADS, Calicut, January.

Kerala Law Department (1975), "The Kerala Scheduled Tribes (Restriction on Transfer of Lands and Restoration of Alienated Lands) Act, 1975 [1]", available at: http://keralalawsect.org/acts/Acts1975/Act31_1975/index.html.

Kerala Law Department (1976), "The Scheduled Castes and Scheduled Tribes Orders (Amendment) Act, 1976", Trivandrum: Government of Kerala.

Kerala Law Department (1999), "The Kerala Restriction on Transfer by and Restoration of Lands to Scheduled Tribes Act, 1999", available at: http://keralalawsect.org/acts/Act1999/Act12_1999/Act.html.

Kerala, Director of Census Operations (2001), "Census 2001 Series-33, Indian Administrative Service", Trivandrum: Director of Census Operations.

Kerala, Collectorate, Wayanad District (2012), "Distribution of Land to Adivasi Communities in Kerala after Muthanga Struggle", State Public Information Officer, Junior Superintendent, Collectorate, Wayanad District, Right to Information Act (2005), October 11.

Kerala, Tribal Resettlement and Development Mission (2012), "Distribution of Land to Adivasi Communities in Kerala after Muthanga Struggle, Under Right to Information Act 2005", Thiruvananthapuram: Tribal Resettlement and Development Mission, Government of Kerala, October 18.

Kjosavik, Darley Jose (2006), "Articulating Identities in the Struggle for Land: The Case of the Indigenous People (Adivasis) of

Highland Kerala", South India, *Montpellier*.

Kunhaman, M. (1985), "The Tribal Economy of Kerala: An Intra-Regional Analysis", *Economic and Political Weekly*, 20(11), March 16.

Kunhaman, M. (1989), "Development of the Tribal Economy", New Delhi: Classical Publishing Company.

Janu, C.K. (2004), "Mother Forest: The Unfinished Story of C.K. Janu", New Delhi: Kali for Women and Women Unlimited.

Janu, C.K, Jaison Chacko and Subhash Gatade (2005), "No Land Even For Burial", *Countercurrents*, March 23.

Janu, C.K. and M. Geethanandan (2003), "The Return to Muthanga", *PUCL Bulletin*, available at: http://www.the/return/muthanga.html.

Janu, C.K. (2010), "My Life and Struggles", Special Session, Paniya Sadas: A Programme Organised for Paniya Community by Centre for Development Studies in collaboration with Montreal University; Shreyas, A Non-governmental Organisation, Sulthan Bathery, Wayanad District, Kerala, March 13-15.

Kumar, Sunil and K.M. Seethi (2014-15), "Marginalisation, Identity Politics and New Social Movements", *Indian Journal of Politics and International Relations*, Vol. 5-8.

Luiz, A.A.D (1962), "Tribes of Kerala", New Delhi: Bharatiya Adimjati Seva Sangh Publishers.

Mathur, P.R.G. (1977), "Tribal Situation in Kerala", Trivandrum: Kerala Historical Society.

Parayil, Govindan (ed.) (2000), "Kerala: The Development Experience, Reflections on Sustainability and Replicability", London: Zed Books.

Prabhakaran, G. (2003), "Road to Muthanga: Sabotaging the Tribal Act", *The Hindu*, April 28.

PUCL (2003), "Statement against Police Firings on Adivasis in Muthanga (Wayanad District, Kerala)", *PUCL*, March 7.

Raman, K. Ravi (2002), "Breaking New Ground: Adivasi and Struggle in Kerala", *Economic and Political Weekly*, March 9, 916-918.

Rammoham, K.T. (2008), "Caste and Landlessness in Kerala: Signals from Chengara", *Economic and Political Weekly,* 43 (37), 14-16.

Rupavath, Ramdas (2009), "Tribal Land Alienation and Political Movements: Socio-Economic Patterns from South India",

United Kingdom: Cambridge Scholars Publishing.

Seethi, K.M. (2001), "Grasim: Polluter Does Not Pay", *Economic and Political Weekly*, Vol. 36, No. 29, July 21.

Seethi, K.M. (2000), "Cleaning the Chaliyar River: Pollution Control or Jobs", *Economic and Political Weekly*, Vol. 35, No. 3, January 15.

Seethi, K.M. (2003), "Adivasis under State Terror", Keralamonitor.com, available at: http://www.scribd.com/doc/4094331/Adivasis-Under-State-Terror.

Seethi, K.M. (2016), "Left Front Victory in Kerala: A Verdict for 'Social Re-engineering", *Economic and Political Weekly,* Vol. 51, Issue 22, May 28.

Shashi, S.S. (ed.) (1995), "Tribes of Kerala", New Delhi: Anmol Publications Ltd.

Sreerekha, M.S. (2010), "Challenges before Kerala's Landless: The Story of Aralam Farm", *Economic and Political Weekly*, XIV (21): May 22.

9

Local Governance and Disaster Risk Reduction

C. Vinodan and Rajeev M.M.

Introduction

Disasters, in any form, affect different people differently. The worst affected in any disaster are the poor and marginalized groups, not only in terms of loss to life, livelihood and habitat but also the loss to their dignity and capacity to recover from the devastation caused. The destruction brought by disasters for women and children is the worst as they are the most vulnerable sections of the society. Their sources for food become meagre; they are rendered homeless by nature and are prone to every form of insecurity. Disasters also have a large impact on the development of any country. Capital channelized into relief aid and reconstruction coupled with the loss of domestic product percolates the country further into extreme poverty and increasing its vulnerabilities to future disasters.

Looking into the figures, it is quite evident that natural disasters are increasing in India and almost all parts of the world, due to numerous reasons such as pollution, environmental destruction, deforestation, population growth, climate change, nuclear outbursts, and lack of effective implementation of environmental and related policies. There is also worldwide concern to mitigate the growing incidence of disaster and their peal on human life, property, and environment.

The role of panchayats is central in sight of their vicinity to the local community, widespread coverage and highlighting people's involvement on an institutionalized basis. It is inevitable to effectively involve panchayats in disaster preparedness and response activities and sensitize the local communities to develop coping mechanism in preparedness and mitigation measures to minimize its damaging effect on life and property of the people. As a first responder to disasters, it is the crucial role of the local government to ensure sustainable disaster risk reduction thereby activating development in the life of the people. The local people should consult for democratic decision-making both pre- and post-disaster phases in all activities like need assessment, allocation of appropriate resources, and implementation of development projects.

Sendai Framework for Disaster Risk Reduction 2015-2030, pointed out that the disaster risk reduction and management depends on coordination mechanisms within and across sectors and with relevant stakeholders at all levels, and it requires full engagement of all State institutions of an executive and legislative nature at national and local levels and a clear articulation of responsibilities across public and private stakeholders, including business and academia, to ensure mutual outreach, partnership, complementarity in roles and accountability and follow-up (UNISDR, 2015).

The risk reduction at the local level depends on good local governance, particularly in political decision-making, formulation of policy, and enforcement relating to land use planning, regulatory controls, zoning, and construction standards. Risk reduction calls for flexibility in the decision-making process and the empowerment of communities, which in turn pushes transparency and good governance. The value and usefulness of an alliance of local governments for disaster risk reduction is that it can help facilitate good governance and create space for stakeholders to work together. The dynamic commitment and facilitation of panchayati raj institutions is important for the implementation of any local disaster risk reduction measure to deal with different stakeholders and multiple layers of government. In many cases, a comprehensive disaster risk reduction measure takes a longer time to fully implement, and the leadership of the local government is particularly crucial to ensure political momentum and support among external stakeholders throughout the process (UNISDR, 2010).

Rights-based approach to development is an approach to development supported by development agencies and civil society organizations to achieve a positive transformation of power relations among the various development actors (Gneiting, Uwe et al., 2009). This practice blurs the distinction between human rights and economic development. There are two stakeholder groups in rights-based development—the rights holders (who do not experience full rights) and the duty bearers (the institutions obligated to fulfil the holders' rights). Rights-based approaches aim at firming the capacity of duty bearers and empower the rights beneficiaries. The author here pointed out some of the experiences gained from various disasters affected places in the country.

The study clearly says that the ultimate responsibility of the local governments in preparing, reducing, and mitigating disaster management activities are very crucial and the post-disaster management efforts such as disaster response and recovery efforts will be effective only if the local government facilitates the process with the cordial participation of other stakeholders. Policy level support, administrative support and community participation are the crucial elements in disaster risk reduction. The following brief literature on DRR underlines the significance of community-based disaster risk reduction and also highlights the rights of the disaster affected community.

Review of Literature

Nevertheless to say that, the local self-government institutions are the most appropriate mechanisms for involving people in natural disaster preparedness and related activities. The ensuing analysis discusses the major roles of PRIs during a crisis situation. The functionaries of the panchayati raj institutions involved in disaster management can train the people to face disasters through community initiatives; keep people in safer places and maintain proper administration of relief in emergencies; mobilize people and sensitize them to have enough courage and preparedness to face the situation; reduce the extent of loss and damages caused by the disaster; take initiative to develop micro plan for reconstruction; and initiate long-term measures to prevent or minimize the intensity of disasters (Jain and Polman, 2003).

It is well written in the 73rd Constitutional Amendment of India about the crucial role of village panchayats—the sole responsible bodies for the development of villages and function as local self-governance institutions. Their primary duties are: (a) effective habitat development including zoning for safe locations and sites for the site less; (b) providing basic infrastructure such as roads, electricity, drainage, waste management, safe drinking water and sanitation facilities; (c) providing adequate employment to ensure food security and to prevent starvation and deaths; and (d) disaster preparedness and management (PRIs, Planning Commission, 2001). Apart from these mandatory responsibilities, there are "29 other areas such as, but not limited to, land and water management, vulnerability reduction, social security, and child development" earmarked as the duties of village panchayats under Section 243G of the Constitution of India.

(Jena, 2009) in his article on 'Empowering Panchayats and Community for Disaster Preparedness and Mitigation: An Institution Building Perspective', has mentioned that panchayats are not consulted for planning in disaster preparedness and relief and rehabilitation work. It would lead to lack of transparency, accountability and social audit in disaster mitigation efforts. There should be introspection as to how the local communities and local institutions like panchayats face such disasters and how best they can handle the crises.

There is a view that local bodies like panchayats should be encouraged and empowered to manage the local affairs with the available local resources. The elected leaders and officials of panchayats are to be trained to develop capacities to handle crisis situation in terms of preparedness, warning, rescue, relief, medical assistance, damage assessment, psychosocial care, water and sanitation and rehabilitation operations. It is felt that even during the largest disasters the role assigned to panchayats is meagre in handling various problems at the bottom level (Sharma, R.K. and Sharma, Gagandeep, 2009).

UNDP reports (2004) underlined about the uncompromised role of the panchayati raj institutions in moving towards sustainable development and successful disaster risk reduction. It is being increasingly recognized that disaster risk management at the local level is a key element in any viable national strategy to reduce disaster risk.

Capacity Building to Reduce Disaster Risk

The ISDR Secretariat of the United Nations defines capacity building as efforts aimed to develop human skills or societal infrastructures within a community or organization needed to reduce the level of risk (UN/ISDR, 2004). In the context of disaster risk reduction, capacity building can be achieved through disaster management training and education, public information on disasters, the transfer, provision or access to technology or other forms of technical assistance intended to improve institutional efficiency. The concept also relates to the training of disaster managers, the transfer of technical expertise, the dissemination of traditional knowledge, strengthening of infrastructure and enhancement of organizational abilities (UN/ISDR, 2004).

Major disaster events in recent years, for example the 2004 Indian Ocean earthquake, have shown the need for greater education and information in disaster risk management. In these relatively poor parts of the world, especially setting up the communications infrastructure to issue early warnings is a big problem. To achieve improvements concerning this goal, the process of capacity development should be addressed at all levels and all sectors (FIG, 2004). In the 21st century, the key issues in capacity building efforts are strengthening of the legal and organizational capabilities of institutions in charge of disaster risk management and networking between them.

The brief literature indicates that the local self-government should take ownership of the community-based disaster risk reduction programmes which should be carried with an active involvement of the community and other stakeholders.

Study Objectives

The objectives of the paper are as under:

1. To highlight the significance of local governments in disaster risk reduction programmes.
2. To find out the role of local governments in understanding local risks and vulnerabilities.
3. To understand various obstacles faced by local bodies in co-ordinating DRR activities within the community.
4. To find out ways to improve the functioning of local governments in addressing disaster risk reduction at the local level.

Materials and Methods

The methods adopted for this study is qualitative and the data collection was done through FGDs, one to one interaction and qualitative interviews. The researcher used purposive sampling method for selecting the samples and the sample size for interview was 20. An interview guide was prepared which contained questions specifically intended to capture the problems encountered by local self-governments in understanding disaster risk and coordinating the DRR activities at the community level.

The researcher administered tools of data collection namely, informed consent form, and interview schedule for qualitative assessment. The investigator obtained informed consent prior to the qualitative and quantitative interviews. The study respondents were asked to complete a brief socio-demographic schedule which gathered details relating to their name, age, gender, ethnicity/caste, religion, education, occupation, income, and type of loss. An interview guide was prepared by the researcher for collecting qualitative details. The qualitative interviews were conducted in regional language by the principal investigator and it lasted for one hour. All the interviews were documented and later transcribed for analysis.

Apart from primary level data collection, the researcher collected resources from secondary data such as books, journals, magazines, articles, newspapers and related sources. The investigator also carried out field visits for data collection and conducted individual and group meetings or interaction with various stakeholders during the study period. The meetings were conducted in various places of Tamil Nadu, Gujarat, Maharashtra and Andaman Nicobar Islands. Qualitative analysis of the data using axial coding revealed the major themes explained in the findings and discussions as part of this paper.

Findings and Discussion

Qualitative Findings: The following section briefly describes the results of qualitative data analysis identified through axial coding.

Researcher identified the following major themes: information building; coordination of pre- and post-disaster initiatives; delegation of power; coordination among various actors; disaster finance; facilitating participatory risk identification and vulnerability analysis exercises; capacity development; identifying the deserving survivors; impact assessment; centralized disaster relief operations; early warning system; utilizing community resources; research and studies etc.

Information Building: A consistent theme that came up across participants' narratives on knowledge building/ information building among the community was that the local self-government (LSG) should take up the lead role in disseminating the disaster and emergency related programmes such as conducting meeting with various stakeholders, imparting capacity building exercises, training to different stakeholders, coordinating the updates of the disaster risk, vulnerability and risk reduction, recruitment of village volunteers, community based resources mapping etc. LSG members must be available when needed, should be trained to transmit knowledge and guidance to the public about accessing resources. Participants felt that district and state government authorities must work through LSG members to provide support and services to disaster prone areas or the affected places for the support of the community.

Co-ordination of Pre- and Post-disaster Initiatives: The coordination from the local governments should start from the pre-disaster initiatives such as disaster prediction, disaster prevention, disaster protection, disaster planning, disaster preparedness, disaster promotion, disaster publicity, disaster participation and disaster partnership. The post-disaster activities, such as disaster rescue, disaster response, disaster relief, disaster rehabilitation, disaster reconstruction, disaster retrofitting, disaster recovery, disaster resilience and disaster research programmes should be facilitated by the bottom level for effective intervention and management.

Delegation of Power: Local government is one among the first responders and is in close proximity to the disaster affected area. Study participants advocated for fundamental policy changes that give more power to local governments in dealing with the crisis. LSGs should have power in making decisions regarding local community's disaster preparedness needs, building local capacity, organizing immediate emergency relief operations, identifying beneficiaries and distributing benefits, planning short-term and long-term recovery goals, and coordinating with the local, national, and international organizations engaged in relief operations.

Local representatives are unclear about the role of local government and are very much apprehensive about the present patriarchal model top-down service delivery approach followed in the management of an emergency. Participants demanded change in the current approach where district authorities gave directions to LSGs to manage relief operations and recovery efforts. According to them, district authorities are not in touch with ground realities unlike LSGs and this causes unnecessary delay in processing relief applications and implementing the relief efforts. In addition, participants argued that the National Disaster Management Act needs to specify the role of LSGs and prepare them with periodical trainings to deal with any crisis.

Disaster Finance—Increased funding for LSG: One of the major challenges experienced by the LSGs was the lack of funding. Funding was crucial in building shelters, building and maintaining the seawall, implementing early warning systems at the ward level, capacity building to ensure better preparedness, and supervise and evaluate recovery efforts. In addition, participants felt that funding was required to arrange for transportation, maintaining relief and recreation centres, and coordinating emergency response in the initial phase. The NDM Act should set aside enough relief funds for disaster prone areas which would help speed up the relief and recovery operations subsequent to any crisis.

Co-ordination among Various Actors: The study indicates that the local community-based organizations such as ward committees, Kudumbasree groups, temples, churches, and other religious groups can play a significant role in disaster management programmes especially in disaster planning, preparedness, relief and recovery efforts. Empowering local communities and promoting positive collaboration between LSGs and local agencies would speed up the recovery process. The various actors can provide support to the community in identifying disaster risks, identifying the most vulnerable places or locations, communities or groups, understand the causes of vulnerability etc. So, understanding, analyzing and coordinating among the community members will build up a sense of ownership among the community members that will obviously boost up the confidence and courage to face any disasters.

Facilitating Participatory Risk Identification and Vulnerability Analysis Exercises: The research study indicates that the local government can initiate and facilitate community-based participatory activities in the disaster risk reduction programmes with the involvement of various community stakeholders in the selected locations. Most of the respondents expressed that the local government should operate as a nodal agency to coordinate the disaster risk reduction activities incorporating various stakeholders and inputs.

Capacity Development: In facilitating effective disaster risk reduction programmes at the grassroots level, the respective local self-government needs to internalize the fundamental aspects of disaster management including both pre- and post-disaster programmes. The ward counsellors and related officials and key persons should be trained in disaster management and for providing disaster management related training programmes targeted at various stakeholders. The community members need to be sensitized in disaster preparedness, mitigation and response programs for the effective intervention of disaster management programmes with the support of other stakeholders. It is indeed significant to understand and analyze the community risks, vulnerability, the coping capacity and disaster resilience of the community. Thus, capacity development activities involving the local government institutions and the community form an integral part of disaster risk reduction programmes.

Identifying the Deserving Survivors and Impact Assessment: The identification of right beneficiaries and the process of impact or damage assessment process is considered as one of the most important activities in the post-disaster phase. Identifying the right beneficiaries and addressing their needs, concerns, and their rights will be one of the major tasks performed by the local authorities. In disaster risk reduction, the responsibility in rightly understanding and addressing the risk locations and most vulnerable populations are exactly the priority of the local government.

Centralized Disaster Risk Reduction Operations; Early Warning System; and Utilizing Community Resources: The study pointed out that the disaster risk reduction programmes must be cartelized under the local government. It means the local self-government can act as a central point of all activities that would greatly enhance the existence and sustainability of the programme. It can be improved by follow-up and monitoring of the activity more efficiently. Most of the respondents opined that the disaster preparedness activities such as setting up of control rooms, establishing early warning system etc. should be taken up by the local government institutions in planning and execution of the activities at the community level. Another notable point that came in the study was the assessment, allocation and utilization of the community resources that can be determined and planned by local governments for an effective and equitable distribution in the disaster risk reduction activities and also in the post-disaster context.

Research and Studies: The respondents expressed that the local government can act as a resource centre for the disaster risk reduction operations and can extend support to various agencies involving research studies and other scientific assignments related to disaster risk reduction and allied subjects. The majority of the respondents opined that the local government should initiate some scientific studies and project proposals with the collaboration of various international, national and state level agencies for the protection of their life and environment from natural disasters and environmental issues.

Recommendations

The recommendations put forward by the investigator are as follows.

Reorganizing Disaster Management: The process of dissemination of disaster warning and risk avoidance action necessarily depends on the creative participation of local people in a disaster-prone area. Hence, panchayats and other local mechanisms should have substantial roles to implement an appropriate action in the targeted places. In fact, planning towards management of natural disasters has to combine both higher and lower level administration, that will help in disaster management exercises. The most significant point is that development plans incorporate disaster mitigation norms, catching early warnings, prediction of natural disasters in time, issue of warnings from the disaster management authorities, availability of necessary articles (such as, search and rescue equipments, high wind-proof radio masts, emergency communication infrastructure, etc.), having an emergency response and contingency plan, activating the emergency response plan promptly during a disaster, and providing a comprehensive as well as an effective relief and rehabilitation to the people affected which could be undertaken with a top down planning approach.

However, several important activities, such as, creating awareness among people, promotion of disaster mitigation and preparedness measures among the local communities, dissemination of warning among people in the affected areas, evacuation of people to safer places in the event of a disaster, and taking precautionary measures in the post-disaster situation (such as against epidemics), etc. need to be undertaken with a bottom up planning approach.

Disaster Management and Development Planning: Under this category, an important aspect is that disaster management needs to be integrated with the normal development planning at all levels. Additionally, relevant departments should allocate a certain percentage of the plan budget to carry out pre-disaster management activities.

Invest More on Pre-disaster Activities: Locally appropriate prevention measures such as bio-shields, seawalls and other measures must be erected to prevent further erosion and disasters. A disaster preparedness plan must be drawn up for action at various levels and must be a part of the Annual Development

Plan. There must be regular disaster preparedness drills at community and institutional levels (Orissa Model) on fixed dates to prepare local communities in disaster preparedness. Government officials must be easily accessible and available on all days (even on holidays) in times of emergency.

Attitudinal Change among Government Officials: It has been witnessed from various disasters in the country that the government authorities have been treating disaster management operations only as an emergency responsibility. The current study has underlined the same and this approach needs to be changed and the authorities must learn to tackle natural disasters as a regular phenomenon. Therefore, for effective handling of natural disasters, some possible strategies and measures should be implemented among the government officials including training programmes and practical workshops related to disaster preparedness and mitigation.

Role of the Community: It is felt that the disasters can be effectively mitigated if local communities and government are adequately prepared to handle them. During the interview with the survivors of Tsunami, the respondents stressed that local communities must be made capable to spearhead any relief and rehabilitation operation in the wake of a disaster in a participatory and cohesive manner. The creation of "hazard-resilient bio-regional communities" with emergency reshuffle plans handled by fully-equipped and well-prepared disaster prevention and preparedness committees of village groups are now the need of the hour.

Psycho-social Care and Development Efforts: Integrating the various components of psycho-social care into the development planning needs to be taken care of in the pre- and post-disaster phases. Experiences from the disasters like Tsunami in 2004, have revealed that the timely requirements of the survivors of disaster should be considered on a higher priority and the implementing agencies must integrate psycho-social care with other services during the disaster management operations.

Facilitating Participatory Risk Identification and Vulnerability Analysis Exercises: By achieving this, the local self-government can utilize various participatory tools like participatory vulnerability analysis, hazard mapping, seasonality calendar, capacity and vulnerability analysis and PRA tools.

Improved Co-ordination among Relief Organizations: One of the major problematic areas during the relief work period is the lack of coordination among civil society organizations and higher authorities and interference of local political party forces. A higher level of duplication of services such as relief kit distribution, allocation of houses and other resources by various organizations were observed during the relief and rehabilitation phase. On the other hand, there was a certain amount of coordination mechanism that was exercised in places like Alappad (Kerala), Kanyakumari (Tamil Nadu) and Port Blair in Andaman and Nicobar Islands. In most of these places, consultative meetings were carried out by the coordinated agencies and they tried their level best to work as a coordinated team for the betterment of the survivors of disasters. These practices need to be developed and strengthened in future so that the service delivery proves fruitful and effective.

Disaster Management in Educational Curriculum: Most of the respondents opined that during the qualitative interview session, disaster management should be included in the educational curriculum which includes both theoretical and practical aspects. By inculcating a culture of preparedness and prevention, this step could prove beneficial for the students as well as other groups like youth, women and the community. For this, mock drills and street plays can be useful in creating awareness and practical wisdom among the stakeholders. This should be considered as one of the most important suggestions derived from the current study.

Ensure Effective Community Participation: The government should actively promote—through its own agencies, NGOs and other stakeholders—active community participation in risk assessment, vulnerability analysis, mitigation, planning and implementation of response and rehabilitation activities.

Task Group Formation: Formation of specialized action group(s) for disaster response and recovery activities could prove useful and can heighten the task of facilitation with support from multiple stakeholders. The task group can be classified into various areas on the basis of requirements such as search and rescue groups, food, water and shelter management, sanitation, first aid, financial resource groups, capacity building, information sharing, security, rapid damage assessment, rehabilitation and reconstruction team, resource mobilization team, environmental protection groups, psycho-social team, research and documentation.

Policy Requirements: The local government should focus more on sensitizing and in ensuring implementation of various administrative and related policies towards construction of houses, buildings, bridges, tourist related constructions, adherence to the coastal zone rules and regulations, etc. with the support of a government order or policy in the disaster prone areas.

Conclusion

It is well-understood that the local government's engagement in disaster risk reduction programmes are ideal to understand when analysing the various types of risks and vulnerability a community is exposed to. Unless and until, there is a comprehensive participation of the local government in designing disaster preparedness, prevention and any disaster risk reduction activities, the whole exercise will not get any momentum and will not be fulfilled. The paper concludes that LSGs can play a heightened role in providing and coordinating relief efforts, planning evacuation, reducing psychological stress due to disasters and formation of disaster task groups for an effective disaster preparedness and response. We hope that the lessons learnt from previous disasters and also from the observations from the various interventions in disaster management in the country will extend better solutions in the management of natural disasters in future.

References

Jain, S.P. and Polman, W. (2003), "A Handbook for Trainers on Participatory Local Development: The Panchayati Raj Model in India", RAP Publication 2003-07; FAO Regional Office for Asia and the Pacific, Bangkok, Thailand.

Jena, A.C. (2009), "Empowering Panchayats and Community for Disaster Preparedness and Mitigation: An Institution Building Perspective", in R.K. Sharma and Gagandeep Sharma (eds.), 'Natural Disaster: Institutional Building for Mitigating the Impact on Vulnerable Sections', A.P.H. Publishing Corporation, New Delhi.

Gneiting, Uwe, Bruno-Van Vijfeijken, Tosca and Schmitz, Hans Peter (2009), "Setting Higher Goals: Rights and Development", *Monday Development*, 27(12): 19-20.

Inter-Agency Secretariat of the International Strategy for Disaster Reduction (UN/ISDR) (2004), "Living with Risk: A Global Review of Disaster Reduction Initiatives", Geneva, UN/ISDR, 3.

ISDR (2007), "Disaster Risk Reduction: A Global Review", consultation edition, prepared for the Global Platform for Disaster Risk Reduction, first session, Geneva, Switzerland.

Planning Commission (2001), "Report of the Task Force on Panchayati Raj Institutions (PRIs)", Government of India, New Delhi.

Sharma, R.K. and Sharma, Gagandeep (eds.) (2009), "Natural Disaster: Institutional Building for Mitigating the Impact on Vulnerable Sections", A.P.H. Publishing Corporation, New Delhi.

The International Federation of Surveyors (2006), "The Contribution of the Surveying Profession to Disaster Risk Management", Frederiksberg, Denmark, FIG Publication No. 38.

UNDP (2004), "Reducing Disaster Risk: A Challenge for Development", New York, United Nations Development Programme, available at:
http://www.undp.org/bcpr/disred/english/publications/publications.htm.

UNISDR (2010), "Disaster Risk Reduction", United Nations,

Geneva.

UNISDR (2015), “Sendai Framework for Disaster Risk Reduction 2015-2030”, United Nations, Geneva.

Wisner, Ben, Blaikie, Piers, Cannon, Terry and Davis, Ian (2004), “At Risk: Natural Hazards, People’s Vulnerability and Disasters”, second edition, Routledge, New York.

10

Re-visiting Gandhian Non-cooperation Strategy

Praveen Kumar

The Non-cooperation Movement symbolises, in many ways, the advent of modern democratic methods, institutions and practices in India's history. Seen from politico-sociological point of view, leadership under M.K. Gandhi and his ability to mobilise the people on a mass scale for the cause of India's freedom from the British colonial rule were the two most significant aspects that would remain the highlight of this paper. The Non-cooperation Movement is not being seen as a single event in itself, but a process that inculcated and instilled the democratic values in the Indian masses. Before making a modest attempt in that direction, it is necessary to acknowledge that the significance of the movement and Gandhi's role has been discussed, debated and researched widely from various aspects by the Indian and foreign scholars from different disciplines. However, the issue and the personality of Gandhi, both can be seen in a new light, each time an attempt is made to revisit "Gandhian philosophy and politics".

The paper would confine itself, in scope, to the analysis of the Non-cooperation Movement by bringing in the salience of the leadership ability of Gandhi and how his politics roped in and introduced certain elements that are considered hallmark of modern democratic practices. In doing so, it would be fruitless to connect his philosophy to any of his consistent thinking that would give a solid linkage to any pre-meditated theoretical foundations, because Gandhi himself had remarked that the only thing that was consistent about his ideas or practices was the pursuit of truth. [1]

However, based on my understanding of the elements that constitute the contemporary thinking on modern democracy, I would try to link the politics as Gandhi had introduced as a practice on the Indian political scene with the Non-Cooperation Movement with the former, in parts, at least. This way, my paper is not going to be chronologically ordered to make it to read like a narration of incidents in sequence and the way they unfolded during the course of the events. The endeavour here is to bring out the deeper and broader processes that were to shape the future course of Indian National Movement and the Indian politics, at large. I, will, however, set the context at first in which the things were unfolding.

Colonial Environment and the Indian Response

The emergence of the Non-Cooperation Movement can be seen in a context in which the educated and the professional class, under the aegis of the Indian National Congress was agitating for certain political and civil rights that would allow them to be treated at par with the Europeans. The leadership was responding to the repressive and discriminatory colonial rule in a limited way and at its own level, even though a section that was known as extremists had also existed. The leader of the extremist section, Bal Gangadhar Tilak had even declared that *Swaraj* was the birth right of Indians. Tilak along with Annie Besant and others were also instrumental for the Home Rule Movement that was joined later by other leaders like Motilal Nehru, Jawaharlal Nehru and M.K. Gandhi who became its president in 1920 (Brown, 1974:276).

The methods adopted, overall, were largely of a kind that required petitioning and submitting prayers to the British authorities for the grant of certain rights or obtaining certain privileges for the target audience, as far as the moderates were concerned. The extremists were militant in approach in a sense

that they emphasised on *Swadeshi* and boycott, and aimed at the end of the British rule. The masses, however, appeared to be largely passive, even as the extremist leadership invoked India's past to arouse the masses or to generate the spirit of nationalism. Amidst this kind of political environment, certain degree of national awakening, of course, had taken place among the Indian masses as regards understanding the adverse political, social and economic impact of the colonial rule and the larger necessity to get rid of colonialism.

The making of the context at that point in time, that is at the beginning of the twentieth century in India's history, thus, can be understood something like this. There were leaders of eminence like Bal Gangadhar Tilak, Lala Lajpat Rai and Bipan Chandra Pal, among others, who were convinced that the British rule should end and to do this, the mass support was required. The mobilisation, however, was limited to a few areas of Punjab, Bengal, Bombay and Madras Presidencies. Many a times, the mobilisation was seen linked to or grounded in Hindu religious practices, which limited somehow the participation of the Muslims. Masses at the national level still did not connect fully with the leaders or the anti-colonial struggle. Secondly, the colonial struggle itself did not witness much direct action from the leadership.

The British, still, considered that the movement could be suppressed through force and other repressive legal or administrative measures. While history took its own course, the events that unfolded on the Indian political scene in the first quarter of the twentieth century provided stimuli to the course of events that would shape the future of anti-colonial struggle and also leave the lasting imprint on the evolution of political India. All that happened under the leadership of M.K. Gandhi. It is not that Gandhi created the movement, but what he did was to give direction to the movement that existed only in a scattered form, appeared divided and emerged only in patches across India.

Three events that can be recounted in this light include the happenings in Amritsar in the first four months of the year 1919—the British government's imposition of what was known as Rowlett Act to curb the freedom of speech and assembly of the Indians and agitation of Indian Muslims under the leadership of Mohammad and Shaukat Ali to press the British to offer favourable terms for the Turkish emperor following the end of World War I. Gandhi's role, with respect to these events, can be seen in terms of what he himself had said, "that the leaders do not create mass movements, but the role of leader is to anticipate the minds of the people and understand when the mass was on move" (Chandra, 2011:19).

The imposition of the Rowlett Act, despite the opposition from the Indian leaders had aggravated the situation. Indians were hoping for a better deal following their cooperation with the British government during World War I. However, the British government had chosen to continue with its wartime policy that allowed the British rulers to arrest without any trial and sustained imposition of censorship on the press. It was against this backdrop that Gandhi gave a call for *hartal* (strike) to be observed with fasting and prayers across India on April 6. Due to 'miscommunication', *hartal* had been observed in Delhi on March 30 during which rioting had also occurred. Gandhi condemned the violence and declared for the withdrawal of *hartal* later on April 18, 1919. Meanwhile, Gandhi had not been allowed to visit Amritsar, where massacre by the British forces had occurred at Jallianwalla Bagh on April 13, 1919.

Following the Jallianwalla Bagh incident, the Indian sentiments against the British were running high and the official Hunter Committee that had been formed to look into the act of Brigadier General Dyer had given a report that was considered to be an eye wash. Louis Fischer has opined that it was the Jallianwalla Bagh incident that had drawn Gandhi into politics (Fischer, 2010:235). The earlier activities were considered moral and socio-religious. Hunter Committee report had absolved the Punjab Governor Michael O' Dwyer. The conclusions of the report were weak. General Dyer was held responsible for the "grave error of judgment". General Dyer had been asked to resign from the Army.

However, the British House of Lords gave Dyer a vote of approval and British admirers presented Dyer with a sword of Honour and 20,000 pounds (Gandhi, 2006:239). The incidents were painful to the Indians. Another action of General Dyer further hurt the sentiments of the Indians. This was known as Dyer's infamous crawling order under which Indians passing by the house of a British teacher who allegedly had been assaulted by Indians had to crawl to cross that street on which the teacher's house was located. Dyer had also got a flogging post erected there for the violators of that order. Gandhi stated that he had felt worse about the 'outrage' than the Jallianwalla Bagh massacre (Fischer, 2010:234).

While these events were adding munitions to the rising Indian sentiments against the British, the Khilafat movement that was being led by Ali brothers gave Gandhi an opportunity to channelise the scattered energies of the masses towards an identified political goal. Gandhi had been invited to a Muslim Conference in Delhi in November 1919. The Conference pertained to the course of action that the Indian Muslims should take in the wake of the terms of the treaty that had been imposed on Turkey following the signing of the Armistice on November 11, 1918 after the end of the World War I. Indian Muslims may not have been guided by a sense of pan Islamism, however, the fate of the Caliphagitated Muslims in India (Ibid:238).

Invitation letter to Gandhi found mention of cow protection, as well, to which Gandhi had suggested to the conference that the two issues should be treated as detached from each other. If Hindus were to support, what would be known as the Khilafat Movement, then this should be done irrespective of Muslim's attitude towards cow slaughter. At the same time, if Muslims were to respect Hindu sentiments regarding cow protection, this should be done without linking it to Hindu's support to the Khilafat Movement. Following Gandhi's advice, cow was removed as an item of discussion from the agenda of the Conference.

It is here, at the November 1919 Muslim Conference that Gandhi had suggested the idea of non-cooperation. He stated that "Indians could not simultaneously oppose the government and work with it" (Ibid:239). As it would come out later, the Treaty of Sevres of May 1920 deprived Turkey of all her colonies and of Greek majority areas. It placed Mecca and Medina under a pro-British Chieftain and gave Britain the guardianship of Iraq that contained Karbala and Najaf and also Palestine that included Jerusalem. "All of Islam's holiest sites were thus placed under the direct or indirect influence of Britain, while France, another non-Muslim country, was asked to 'advise and assist Syria" (Gandhi, 2006:238-239). The terms of the treaty only hastened the launching of the Non-Cooperation Movement.

Gandhi saw in the Khilafat Movement an opportunity to cement Hindu-Muslim unity that the British Raj had successfully prevented, to sustain its rule. According to Rajmohan Gandhi, the period of Non-Cooperation Movement saw, for the first time, since 1857, a united front of Hindus and Muslims (Ibid). However, as per some other scholars, Gandhi's support to Khilafat was influenced by his deeply religious outlook. B.R. Nanda opines that Gandhi understood Khilafat Movement as largely derived from Muslim divines and pan-Islamist enthusiasts. This point of view has been contradicted by Louis Fischer, as stated above.

In Nanda's view, Gandhi "failed to see that the Khilafat was a moribund institution, that the Turks themselves were sick of it, that Ottoman empire could no more remain intact after the War than the Hapsburg Empire, and that the smaller nations, Arab and non-Arab, were struggling to be free from the stranglehold of Turkey (Nanda, 1958:185). The relevance of the institutions of the Caliph and its value could be a matter of debate, however, as would be argued in the following pages,

Gandhi captured well the mood of Indian Muslims and the symbolism that they attached to the institution of Caliphate.

Non-cooperation as Defined by Gandhi

Non-cooperation, according to Gandhi, is not a passive state. It is an intensively active state. It is more active than physical resistance or violence. Gandhi disapproved the use of the term passive resistance for non-cooperation. As described by Gandhi himself, "Non-cooperation, in the sense used by me, must be non-violent and, therefore, neither punitive nor vindictive nor based on malice, ill-will or hatred" (Hingorani, 1998:41). [2] Non-violence remained a cardinal creed throughout in the Gandhian politics and philosophy. This was a vital part of non-cooperation, as Gandhi underlined it while writing in the *Young India* on July 19, 1921. To Gandhi, non-violence meant reliance on God. Hence, the non-cooperator strived to compel attention and to set an example not by his violence, but by his unobtrusive humility.

While writing in the March 23, 1921 issue of *Young India*, Gandhi stated that "Non-Cooperation predominantly implies withdrawing co-operation from the State that in the non-cooperator's view has become corrupt and excludes civil disobedience of the fierce type...". However, "Non-cooperation, too, like Civil Disobedience is a branch of *Satyagraha*, which includes all non-violent resistance for the vindication of Truth" (Ibid:35-36).

As would be explained in the paragraphs below, Gandhi withdrew the Non-Cooperation Movement following incidents of violence in Chauri Chaura in early 1922. Later, Gandhi outlined the lessons that could be drawn from Non-Cooperation Movement. The first lesson, according to him was reasoned and willing obedience to the laws of the State. The second lesson, according to him, was tolerance. He underlined that "we must tolerate many laws of the State even when they are inconvenient" (Ibid:43). Disobedience may become a compulsion, however, when the laws are unworthy of tolerance and immoral. The third lesson according to Gandhi was suffering. This may even require sacrificing property and even, at times, family.

Much later, while writing in *The Harijan* on August 18, 1940, Gandhi underlined that Non-Cooperation was of universal value. Well applied, its use in politics can wholly displace the use of barbarous weapons of mutual destruction (Ibid:35-36 and 45). Gandhi also cautioned that Non-Cooperation, "when its limitations are not recognised, becomes a licence instead of being a duty, therefore, becomes a crime". The dividing line between right and wrong was thin and would become indistinguishable. The line is breakable and unmistakable. He suggested that the true non-cooperator would be doing it for the right cause and willingly.

The Khilafat Non-cooperation Movement had begun on August 1, 1920, even before the Calcutta session of the Congress in September 1920 that discussed the Non-Cooperation programme in detail and adopted the resolution in favour of the programme with a great majority. There was phenomenal growth of support for the non-cooperation movement within the four months between the Calcutta and Nagpur Congress in December 1920, which adopted the new constitution of the Congress and declared *Swaraj* as the goal of the Congress (Desai, 2009:205). The eventual objective of the Non-Cooperation Movement was to attain *Swaraj*, which Gandhi had declared at the Calcutta session earlier that *Swaraj* could be achieved in one year if the non-cooperators non-cooperated non-violently and with the principles that was outlined for the non-cooperators. Untruth and injustice had to be opposed with dignity. Gandhi understood that training and self-sacrifice were the two aspects of the struggle. Gandhi as a leader had the vision of *Swaraj*, which could be attained only through non-violent means. His vision was love for all. It was a process, means as well as end (Anand, 1999:51).

Democratising the Polity in a Colonial Environment

Non-cooperation Movement could be seen as the first

mass-scale movement that challenged the might of the British colonial powers in an organised manner through the innovative 'non-violent' means almost across India. The movement itself, however, could be seen as something that initiated the process of instilling the values of democracy in a society that was highly stratified and diverse. The anti-colonial movement itself had lacked mass scale participation and uniform spread across India. What the Indian anti-colonial struggle required at this stage was unification of various segments, mainly Hindus and Muslims as two different categories, on a unified platform to fight the oppressive British colonial rule.

The second aspect that was significant, but was lacking in the Indian freedom movement at that time was the connectedness that should exist between the movement, the leaders of the movement and the masses. In other words, the movement lacked the character of being something, which existed for the masses and which was aiming to achieve *Swaraj* for the masses. The one element that was being considered as given at this point of time is that the transformation of the society in the historical stage next to colonialism in India was taking place towards democratisation and not towards any other kind of political system. Democracy, because this may be not the best form of government that we have, but because no other alternative form of political organisation in history is known to have cherished better values than democracy.

Now, my endeavour in the following paragraphs is to outline some of the values that are celebrated in any democratic form of political organisation and how Gandhi instilled or introduced those elements as leader in the Indian National Congress and the Congress-led National Movement. The elements or values that are being discussed below are not necessarily in that order. The one element that is important in any form of political organisation is that the followers should remain attached to the leader. In my view, the leader's *charisma* becomes an important element in widening the acceptability of the leaders among the followers. This is all the more important with respect to democracy that the masses should trust the leaders.

By charisma, here I employ Max Weber's definition that understands charisma as "certain quality of an individual personality by virtue of which he is set apart from ordinary men and treated as endowed with supernatural, superhuman, or at least specifically exceptional powers or qualities" (Weber, 1947:358-359). Thus, according to Weber, the charisma gives the leader to be personally recognised as "the innerly called leader of man". The leader is not obeyed because of some statute or law or because the leader has been obeyed traditionally, but because "men believe in him" (Girth and Mills, 1946:79).

This is something that could be seen very clearly in Gandhi's personality. People of India believed his personality. They trusted and obeyed him. Shahid Amin in his work has brought out what he states that is not the charisma of Gandhi *per se*, but "how this registered in peasant consciousness" (Amin, 1984:289). Amin also notes that most of the rumours about *mahatma's pratap* (power/glory) were reported in the local press between February and May 1921. These related to miracles that invocation of Gandhi's name could bring or the punishment that could come to a person defying the 'authority' of Gandhi. For instance, Amin cites an editorial of April 23, 1921 in the *Pioneer*, which had referred to the stories brought out by the *Swadesh*, a local newspaper in Gorakhpur district of the then United Provinces.

The stories that were linked to Gandhi were, for instance, about smoke coming out of well and scented water of the well, a person who refused alms to a person begging in the name of Mahatma lost gur (jiggery) and two of his buffaloes, and a Brahmin who defied Mahatma's authority went mad and could be cured after three days only after invoking Gandhi, among others. Amin has analysed fifty such stories and presented in a context that links it to "elite manipulation and control" (Ibid: 292). However, the stories could also be seen as part of

popular imagination and how the masses came to attach super natural and super human powers to Gandhi, even if there may not have been any conscious effort on the part of Gandhi or his followers to spread such 'rumours'.

The other important element, which is considered an important value of modern democratic institutions and that Gandhi incorporated during the Non-Cooperation Movement is 'inclusiveness'. In fact, as Iris Marion Young has underlined, democracies are still criticised for "being dominated by groups or elites that have unequal influence over decisions, while others are excluded or marginalised from any significant influence over the policy making process and its outcomes" (Young, 2000:11). The Movement till the 1910s was seen as dominated by groups or elites that were seen as separate and different from the masses, at large. Seen in terms of Gandhian politics and philosophy, the Non-cooperation Movement under Gandhi's leadership drawn in the diverse segments of society, including women, Sikhs, lower castes and dalits (untouchables) and Muslims on one platform to fight against the British colonial rule.

Khilafat Movement had provided a good opportunity, as discussed above for the Hindus and Muslims to come together against the common enemy. The constructive programme during the Non-Cooperation Movement and subsequently, also meant fight against untouchability. Other social issues like opium addiction and prostitution were also paid attention to during the movement. Gandhi sent a message out to the leaders and workers of Congress that untouchability had to be fought against. He interacted with the people from the lower castes, which were considered as untouchables and on one occasion when he was addressing their gathering, he pointed out that if he were to understand their sufferings he had to be borne as one of them (Gandhi, 2006:236). Gandhi also favoured women's participation in the National Movement and there was a great upsurge in women's participation during the Non-Cooperation Movement (Taneja, 2005:53). The whole of the masses, including the *tribals* in many parts of India, thus, were directed towards anti-colonial Movement. [3]

The next thing that was important as regards Gandhi's role is that he directed the masses towards the stated political goal that was *Swaraj* or self-rule. After becoming the leader of Home Rule Movement, Gandhi declared that the self-rule was to come irrespective of the relations the Indians may have with the British. Gandhi, however, was also aware that to carry on the movement of the non-violent kind that he had envisaged, the people needed to be disciplined and organised. In addition, to sustain the movement and challenge the mighty British Raj effectively, alternative institutions and organisational structure were required. All these would also require sufficient funds. To create effective organisational structure, Gandhi at the Nagpur session in December 1920 suggested a new constitution for the Congress. Gandhi converted Congress into a democratic mass organisation. Congress was to be re-organised with village units, city/district units and with provincial sections. An All India Congress Committee (AICC) of 350 members was constituted to make the policies and run the affairs of the Congress in an organised manner. The Congress was also to have a working or Executive Committee of fifteen members (Chandra et al., 1988:186).

The changes in organisational structure were accompanied by broadening the mass base of the Congress. To do this, Gandhi set an objective of enlisting one crore members in the Congress fold. Six million members or sixty lakh members had joined Congress. The membership fee was reduced to four annas to broaden its base among the poor. This also helped raise funds for Congress activities. Gandhi had envisaged Tilak memorial fund following Tilak's demise on July 31, 1920, a day before the formal launching of the Non-Cooperation Movement. The target of one crore rupee was achieved and the fund was, in fact, over-subscribed (Gandhi, 2006:252 and 254).

The eventual objective was to prove it to the Indians, the

world and to the British colonial rulers, in particular that the latter's rule in India did not have any legitimate or moral basis. The British presence as rulers in India was solely based on brute application of force against approximately 318 million Indians who were opposed to the British presence and were agitating peacefully and non-violently for the exit of the English. At the same time, Gandhi was very clear in demonstrating the legitimacy of the movement and people's support for his non-violent politics and philosophy of Satyagraha. This was the reason that Gandhi would send written communication to the British Viceroy, Lord Chelmsford before launching the movement and this he did before the launching of the Non-Cooperation Movement, as well. He wrote to the viceroy, "I have advised my Muslim friends to withdraw their support from your Excellency's Government and advised the Hindus to join them". The Viceroy had reacted by saying that Non-Cooperation was 'the most foolish of all foolish schemes' (Fischer, 2010:242).

Gandhi had the confidence of the masses and Gandhi had confidence in himself and the support of the masses. In addition to charisma, the sound basis of his philosophy and politics in morality and truth, Gandhi's understanding of the pulse of the masses and his conviction that as a leader he should strike the right chord at right time were also significant in channelising the energies of the masses in the right direction. Two things were significant in this light, and these complemented his geographical re-organisation of Congress and reducing the membership fees of the Congress. These were the method of communication and his appearance as a leader from among the masses.

Hitherto, English used to be the preferred medium of communication of the Congress leadership and the appearance of the Congress leaders would give them an anglicised outlook. Gandhi's adoption of simplicity in appearance and adoption of lion-cloth (dhoti) as his attire, in addition to being grounded in his moral philosophy, also helped him connect with the masses. He would appear as one of them with whom the masses could relate to. At the same time, regional re-organisation facilitated communication in other languages and Gandhi, in addition to the use of English where it was necessary also adopted the language of the masses, which was Hindustani. This helped him communicate better and convey to the masses in a language that the people could understand. Gandhi travelled intensively to reach out to the masses and communicated with them, talked to them and incorporated them at a large scale in the Movement. Gandhi would also respond to all the communication directed to him and would also reach out to the people through his writings in the print media like the *Young India* and the *Harijan*.

Gandhi was also convinced that as a leader, simply opposing to the British policies and rule was not enough. Indians should demonstrate to the world that they were capable of self-rule as per any democratic standards of the time. He was also convinced that for this, alternative policies and methods of governance had to be put in place and the population at large needed to be trained in self-rule and other practices that might be required for self-rule. For this, Gandhi adopted a method that would attack the symbols of British rule and at the same time also come up with the alternative structures that potentially could replace those institutions. For the former, Gandhi outlined a four stage strategy for Non-Cooperation in May 5, 1920 article in the *Young India*.

In the first stage, Indians should return titles and honorary posts. Later, when leaders gave the word, Indians should think of quitting civilian jobs with the government. The third and fourth stages would involve withdrawal from police and military, and non-payment of taxes. Gandhi himself returned to the Viceroy his two South African War Medals and his Kaiser-e-Hind Gold Medal for humanitarian work in South Africa. In the accompanying letter Gandhi said, "I can retain neither respect nor affection for a government, which has been moving from wrong to wrong in order to defend its immorality". [4] He appealed to the Indians to stop using foreign goods,

including liquor and cloths, which hit the British economy. [5] Burning of the foreign cloths in publicly organised bonfires also had symbolism attached to it. The Indians were burning what was not theirs. Burning also meant detaching oneself emotionally and showing to the world that there was no affection with something that might have been foreign.

At the same time, he motivated the masses to adopt *charakha* spinning to produce Khadi cloths. This was a constructive programme which engaged the masses, and at the same time also kept the supply of clothes going so that there was no scarcity of clothes. This would also generate revenue for those who were engaged in *charkha* spinning, and also for the programme.

Meanwhile, Mohhamad Ali Jinnah and Gandhi disagreed on the issue of nature of *Swaraj*. Gandhi changed the name of the Home Rule to *Swaraj Sabha* and the Constitution was also changed to incorporate the element that *Swaraj* need not require connection with the British Empire. Voting to new legislature was boycotted. Sidelined at the Nagpur Congress, Jinnah left Congress at the Nagpur session. Jinnah wanted to continue links with the British Raj for the protection of Muslims against the Hindus in continuation with Sir Syed Ahmed Khan's thinking.

Meanwhile, as part of the constructive programme and positive agenda, hundreds of thousands of *charkhas* were operating. Students boycotted government schools and colleges and nationalists started and established national schools and colleges. Top lawyers Chittaranjan Das, Dr. RajendraPrasad, Motilal and Jawaharlal Nehru, among others ended their practices. Leaders like Subhash Chandra Bose and Jaiprakash Narayan entered the scene that infused new energies among the youth and was also an indication of popularity of the Movement among the youth to the extent that the youth preferred the movement to their promising careers.

In early October 1921, Gandhi and his close colleagues declared severing of the ties of civilians and soldiers with the

British as duty. Boycott of the November 1921 tour of Prince of Wales was also announced. Climax was to be the 'mass civil disobedience' in the form of non-payment of taxes. Based on Sardar Vallabh Bhai Patel's suggestion, Bardoli was selected as the site of mass civil disobedience. Bardoli was selected because many of the people in Bardoli had returned from South Africa where they had seen Gandhi's *Satyagraha*. Thus, as Gandhi would have emphasised, many people in Bardoli could be seen as having some training in *Satyagraha*. Bardoli also lied in Surat taluka that was the location of the first British East India Company factory. There was symbolism attached to it with the decision to take up the mass civil disobedience to challenge to might of the British Empire in Bardoli (Gandhi, 2006:259).

Thus, the whole Non-Cooperation Movement was the manifestation of Gandhian politics and philosophy that was with a vision, which included the twin objectives of ouster of the British rule, without turning the British into enemies of the Indian people and establishment of *Swaraj* with Indians trained in self-rule. All these were to happen non-violently. However, the course of the events that followed also demonstrated that at different stages in the progression of human history, the leader and his or her vision cannot mechanically control the turn of the events. This happened to Gandhi, his philosophy and politics and his leadership, as well. Gandhi was forced to withdraw the movement following instances of violence. An analysis of the decision behind withdrawal of the Movement would follow in following paragraphs. However, this needs underlining here that Gandhi himself had remarked at the instances of violence during *hartal* that had been organised to oppose the Rowlett Act that he had made a 'Himalayan Miscalculation' (Gandhi, 1956:182).

Withdrawal of the Movement: An Analysis

Non-cooperation Movement was withdrawn by Gandhi ahead of the mass campaign that he had planned in Bardoli on February 12, 1922. The withdrawal was prompted by the incident of violence in Chauri Chaura on February 5, 1922. Formal withdrawal became effective from February 12, 1922 after Gandhi persuaded the Congress Working Committee to ratify his decision. As Gandhi reported later in the *Young India* on February 16, 1922, the agitators had killed 22 policemen who had opened fire on a mob that had come for the rescue of some of their colleagues participating in the anti-government procession apparently being harassed by those policemen. [6] Withdrawal of the Non-Cooperation Movement was somehow bound to happen. One, Gandhi had declared at the beginning of the movement itself that in case violence occurred, he would be compelled to call off the movement.

Non-cooperation was called by Gandhi a more potent weapon than violence. Writing in the *Navjivan* on May 16, 1920, Gandhi had announced that he would leave the movement if any violence occurred by any one of non-cooperators or at their instance. Second, there were other developments that were indicative of an eventual withdrawal of the Movement, as would be mentioned in the paragraphs below. Finally, Gandhi as a leader also perhaps understood the pulse of the masses, again, that to sustain the movement was becoming difficult for the masses at a large scale. The masses were showing signs of fatigue and impatience, and as a leader this could have been also a strategic move to withdraw the movement. Thus, this was not an to end to the larger campaign *per se*, but perhaps to retreat to allow the people to regroup and come back with rejuvenated energies. As Rajgopalachari had remarked, "The nation is too weak, too far gone in economic misery to be able to fight and win in one campaign. We have to carry on many campaigns before we can reach our goal" (Gandhi, 2006:273).

So, what were the signs of the fatigue that Gandhi as a leader could have seen, in addition to the mass impulsion to turn violent. Riots, for instance had also occurred in Bombay during the visit of Prince of Wales (later Edward VIII). Congress had declared to boycott the visit of the Prince of

Wales. Riots had followed in which Parsis, Christians and Anglo-Indians became special targets (Chandra et al., 1988: 189). Gandhi had asked the CWC to form an All India Volunteers Corps and the Hindustani Seva Dal for controlling demonstrations. The most important was possibly the decline in the trust level of Hindus and Muslims. Any sensible leader could have sensed the atmosphere of violence that was looming large. Many Muslims were attracted towards *Hijarat*. *Hijarat* meant moving from defiled land towards purer land. A land ruled by a race that destroyed Khilafat was considered defiled, which in the present case was India.

In the spring of 1920, many Indian Muslims, mostly Pakhtuns from North Western Frontier Province moved into Afghanistan. After many casualties and few months' of privation they returned. Gandhi had advised against *Hijarat* (Gandhi, 2006:242 and 244). Meanwhile, the developments in Turkey were indicative of the declining interest of the Turks themselves in the institutions of the Caliph. Mustafa Kamal's rise to power and his disregard for the Turkey Sultan came as setbacks for the Khilafat movement. Meanwhile, charged and emotional speeches also came from Ali brothers, who however, apologised at Gandhi's instance. While association of the leaders continued, but there were suspicions that were emerging at the local level between the two communities. Each community rumoured that the other community professed non-cooperation, but retained government jobs.

Malabar episode of August 1921 further eroded the trust level between the two communities. Moplah tenants had risen to violent rebellion over alleged insult to their religious leaders against the government first, and later against their landlords who were Hindus. Independence was declared and violence occurred. Some Hindus were forcibly converted (Ibid:255). The British Raj took military action and 2,000, mostly Moplas were killed and over 2,400 were convicted of rebellion and other lesser crime. The rebellion was local in origin and aim. However, the British rulers linked it to the Khilafat movement

of Ali brothers. Gandhi had maintained that the British rulers did not allow Non-cooperators to Malabar. The British did not want to give the non-cooperators an opportunity to end the trouble through peaceful means. They were "desirous of showing once more that it is only the British soldier who can maintain peace in India" (Ibid:256). Ali brothers were arrested in September 1921 and Gandhi had strongly reacted to their arrests in the October 29, 1921 edition of the *Young India*.

Many Congress leaders, including Motilal and Jawaharlal Nehru, C.R. Das, and Subhash Chandra Bose, among others were critical of Gandhi's decision to withdraw the Non-Cooperation Movement. Some commentators had also criticised the withdrawal on the ground that the Chauri Chaura incident was an indication of increasing militancy among the Indian masses and Gandhi was wary that this could pose a threat to the propertied class in India. [7] Bipan Chandra has opined that Gandhi's critics were less than fair to him while promoting this kind or argument (Chandra et al., 1988:192). As has been mentioned in the preceding paragraphs, Gandhi had warned at the beginning of the movement itself that any violence would lead to the cancellation of the Movement. Fear of violence was not simply confined to the period of the Movement. As Gandhi himself said, following the Chauri Chaura incident, "Suppose the non-violent disobedience of Bardoli was permitted by God to succeed and the Government had abdicated in favour of the victors of Bardoli, who would control the unruly elements that must be expected to perpetuate inhumanity upon due provocation?" [8]

Gandhi, here, perhaps was indicating to the already declining trust between the Hindus and Muslims. The incidents of violence would be directed against each other that would lead to killings and massacres of the innocent people. Gandhi, as a true leader, here was not thinking in terms of simply throwing the British out of India, but he was also aware that there has to be alternative means to carry on the task of governance in the interest of whole of India after the British would have left. Physical security of an ordinary Indian might have been at the top of his mind. The same also possibly would have applied to prevent any likely repression from the British Government on the whole of India on the pretext of a few violent incidents here and there.

Thus, Gandhi as a leader must have thought that the time was right to retreat in a dignified manner, before things may go out of hand. Gandhi might also have sensed that in many parts of the country the movement was possibly already receding. Students had started returning to the schools and colleges, lawyers and litigants had started returning to the court, the commercial classes might have been wary of losses in the business and the people serving in the government appeared restless.

The attendance at the meetings and rallies was also getting thinner. In addition, as historians inform us "mass movements have an inherent tendency to ebb after reaching a certain height, that the capacity of the masses to withstand repression, endure suffering and make sacrifices is not unlimited, that a time comes when breathing space is required to consolidate, recuperate, and gather strength for the next round of struggle, and that, therefore, withdrawal or shift to a phase of non-confrontation is an inherent part of a strategy of political action that is based on the masses" (Ibid:194-195). The Movement could have been launched at right moment again when the people of India were ready for the Movement. This is what the history of the Indian National Movement witnessed in its next phase that was the launching of the Civil Disobedience Movement under Gandhi's leadership.

Gandhi and Lessons from Non-cooperation Still Relevant

The relevance of Non-cooperation Movement and Gandhi's philosophy and politics in that context can be seen by situating the debate historically and with respect to today's world democratic polity, as well. At one level, the anti-colonial movement acquired the mass character and the masses at an all-India level could connect and relate themselves with the Movement. This was the strong foundation on which Gandhi and Congress leaders further shaped their strategies of Freedom Movement that was based on the masses. This itself had gone a long way in India's nation-building process and could be considered a significant landmark in the consolidation of India's identity. At another level, and this was again a significant contribution in the nation-building process, the modern democratic values that took root during the Movement had shaped the future Indian polity and here the Gandhian philosophy and politics can be taken out of the context of National Movement to see its relevance.

At the advent of the Non-cooperation Movement, Indian society and the National Movement, both had lacked the democratic character. The leadership was more seen as exhibiting elite characteristics and fighting for the gentleman's cause of more participation in administration and equality with the British elite. This changed with Gandhi coming on the scene. This was rather a complex construct where the Congress under Gandhi was organising the masses according to and for the democratic values under the colonial rule, which was inherently undemocratic and repressive. The principles and values that Gandhi adopted to do this continue to remain relevant today.

The principle of non-violence provided a means to fight the enemy to bring to one's side. Gandhi was clear that self-rule had to be won in such a manner that would not lead to loss of life and property of anyone. This was rather an idealistic position, but the underlining philosophy was that the cycle of violence once started could be unending and only the parties involved in violent acts may keep changing. Hatred and violence, once they take deep root, it is hard to eradicate them. The practical application of this philosophy may be seen as having limitations, especially seen with respect to human nature and in the field of statecraft, but as an ideal and philosophy that envisions a secure world, the cult of non-violence would appear to be based on a sound logic.

At the same time, the values that Gandhi intended to achieve and instil among the masses continue to remain significant for a diverse society like India, and also for the rest of the world. The first thing that can be listed in this category is peaceful co-existence. All the segments of the society should live in harmony. Hindus and Muslims should remain not as two different categories, but as equal partners of the same society. This aspect was later reflected in India's Constitution as well, which has given secularism an important place. Peaceful co-existence, in fact, continues to remain equally valid for the international community in today's world, as well.

Needless to underline that in today's world where terrorism that is based on the method of mindless and indiscriminate violence mostly against the innocent civilians continues to remain one of the biggest challenges for India and the world—the creed of non-violence can bring peace and development for the humankind. Equality is another aspect that was emphasised by Gandhi during the Non-cooperation Movement. By bringing in the different segments of the society, including the 'untouchables' and women, he possibly intended to demonstrate that they all have an equal stake and hence responsibility in nation-building. Equality continues to remain one of the basic values of a democratic society, and history is witness to this fact that inequality and the structures that abet and sustain inequality lead to resentment among various sections and can lead to violence and possible confrontations and conflicts among various sections of the society.

Many aspects of the relevance of Gandhian politics and philosophy, especially with reference to the Non-Cooperation Movement have been dealt with in the preceding paragraphs. One, and that possibly I would highlight as the most important aspect is the leadership qualities that Gandhi exhibited during the Non-Cooperation Movement. A leader should be living for the people and should belong to the people. This is the lesson that every leader in a democratic polity needs to learn. People follow the leader and the leader needs to have certain qualities that sets him apart from the masses. It is the leader's ability to lead, provide direction to the people, channelise their energies effectively in the right direction, fight for the cause of the masses and, and above all, lead the movement with responsibility while ensuring security of his followers that continue to remain and would continue to remain as important virtues of a leader in a democratic polity.

Gandhi also taught us that a principled stand and living by the principles are other cardinal virtues that a leader needs to exhibit to the masses.

Endnotes

1. Bipan Chandra has observed that Gandhi seldom wrote theoretically, outside the immediate problem or context. See Chandra, Bipan (2011), "Essays on Indian Nationalism", New Delhi: HarAnand, 18.
2. Gandhi described this aspect of non-cooperation while writing in the *Young India* on August 25, 1920. Quoted in Hingorani, Anand T. (ed.) (1998), "Gandhi for the 21st Century: The Science of Satyagraha", Vol. 4, Mumbai: Bhartiya Vidya Bhavan, third edition, 41.
3. Rajmohan Gandhi writes that tribal stirs led by local leaders were inspired by Gandhi. In his words, "These stirs were not always non-violent or supported by Gandhi himself, but associates of Gandhi (including Indulal Yagnik, Amritlal Thakkar and Jugatram Dave) had by this time started work on ashrams for tribals in southern Gujarat". See Gandhi, Rajmohan (2006), "Mohandas: A True Story of a Man, His People and an Empire, New Delhi: Penguin Publications, 254.
4. Quoted in Fischer, Louis (2010), "The Life of Mahatma Gandhi", London: Harper Collins, 244.
5. As per one estimate the value of imports of foreign cloths alone fell from ₹ 102 crore in 1920-21 to ₹ 57 crore in 1921-22. Chandra, Bipan et al. (1988), "India's Struggle for Independence", New Delhi: Penguin Books, 188.
6. Quoted in Fischer, Louis, Ibid., 252.
7. R.P. Dutt has taken this position and his is a Marxist point of

view. Dutt, R.P. (1940), "India Today", London: Victor Gollancz, 324-329.
8. Quoted in Fischer, Louis, Ibid., 252.

References

Amin, Shahid (1984), "Gandhi as Mahatma: Gorakhpur District, Eastern UP, 1921-22", in Guha, Ranjit (ed.), 'Subaltern Studies III: Writings on South Asian History and Society', New Delhi: OUP.

Anand, Jana (1999), "Gandhi's Leadership: The Oceanic Circle beyond Time and Geography", Mumbai: Bhartiya Vidya Bhavan.

Brown, Judith M. (1974), "Gandhi's Rise to Power: Indian Politics 1915-1922", Cambridge: Cambridge University Press.

"Census Data: 1921", Census of India, available at: http://www.censusindia.gov.in/Census_And_You/old_report/census_1921.aspx.

Chandra, Bipan (2011), "Essays on Indian Nationalism", New Delhi: HarAnand.

Chandra, Bipan et al. (1988), "India's Struggle for Independence", New Delhi: Penguin Books.

Desai, Narayan (ed.) (2009), "My Life is My Message", Vol. II, Satyagraha (1915-1930), New Delhi: Orient BlackSwan.

Dutt, R.P. (1940), "India Today", London: Victor Gollancz.

Fischer, Louis (2010), "The Life of Mahatma Gandhi", London: Harper Collins.

Gandhi, M.K. (1956), "A Himalayan Miscalculation", in Jack, Homer A. (ed.), 'The Gandhi Reader: A Sourcebook of His Life and Writings', New York: Indiana University Press.

Gandhi, Rajmohan (2006), "Mohandas: A True Story of a Man, His People and an Empire", New Delhi: Penguin Publications.

Gerth, H.H. and C. Wright Mills (ed.) (1964), "From Max Weber: Essays in Sociology", New York: OUP.

Hingorani, Anand T. (ed.) (1998), "Gandhi for the 21st Century: The Science of Satyagraha", Vol. 4, Mumbai: Bhartiya Vidya Bhavan.

Nanda, B.R. (1958), "Mahatma Gandhi: A Biography", New Delhi: Oxford University Press.

Taneja, Anup (2005), "Gandhi, Women, and National Movement, 1920-47", New Delhi: HarAnand Publications.

Weber, Max (1947), "The Theory of Social and Economic Organisation", New York: The Free Press and the Falcon's Bring Press.

Young, Iris Marion (2000), "Inclusion and Democracy", Clarendon: Oxford University Press.

11

NGOs and Protection of Human Rights in India

Sijo K. Manuel

Introduction

The NGOs are *non-governmental*, *non-profit-making*, *voluntary*, *altruistic*, *philanthropic* and *private* organisations. The main purpose of these organisations is to provide the necessary services and relief to the socially and economically weaker sections of the society. These non-profit voluntary groups are usually established at local, national as well as international levels. They perform different tasks for solving the problems that enhance the development of the society. The NGOs are connected with government or private sector firms. The social issues which they deal with include women empowerment, gender issues, education, pollution, health, slum dwellers, street children, urban development, human rights, concerns of less privileged etc. They take care to bring up people's concerns and issues to the government and policy makers.

The NGOs fulfil a variety of roles in the society that, among others, include delivering aid and assistance, monitoring, education, grassroots political action and service delivery. They are integral in promoting hope within the society by bringing about a socio economic change. This is a role that is immeasurable, unquantifiable, and intangible. Voluntary organisation is not a new phenomenon in the world. Voluntary effort has always been an important part of our culture and social tradition. The need for organising people into accredited associations and their involvement and participation in rural development have now been fully recognised (Kurakose, 1989). In recent years, they have increased in considerable number, acquired greater importance and put up many new experiments in the field of rural development. The NGOs can play a crucial role by bringing about the socio-economic development of a nation by supplementing the government efforts as they are close to the hearts of the rural people.

World Bank has defined NGOs as private organizations that pursue activities to relieve suffering, promote the interests of the poor, protect the environment or undertake community development work. Another definition is that NGOs are a wide variety of groups and institutions that are entirely or largely independent of governments and characterized primarily by humanitarian or co-operative, rather than commercial objectives. The last decode of the 20th century may be described as the decade of the NGOs since there has been such a massive growth of NGOs in this decade.

The upsurge of NGOs is a worldwide phenomenon, although it has become particularly conspicuous in the less developed countries of Asia, Africa and Latin America. In some of these countries NGOs are seen as the main driving force in economic development and social change. NGOs are now active in every field like literacy, education, health, human rights protection, sanitation, child care, and environment and so on. India is no exception for the growth and development of NGOs in various arenas.

NGOs in India

India has had a great tradition in the realms of philanthropic activities, social service and voluntary work. The instinct of philanthropy is inherent in our individuals as a kind of Dharma (duty) in the Indian philosophy and culture, dating back to the ancient times. It is notable that a large number of charitable and voluntary institutions have emerged in India in the last two centuries to help the poor, the destitute, the downtrodden, the handicapped, and the weaker sections of the society. The voluntary organizations that are engaged in social welfare activities like helping the poor, spreading literacy and so on, have a credible record of achievement in India since the British times (Kothari, 1987). The country's movement for national independence under the Gandhian leadership was rooted in the ideal of social reconstruction, self-help and the upliftment of the poor, especially the untouchables, through voluntary action.

From an operational point of view, the term NGOs has been used in our country to denote those organizations which are non-governmental, voluntary and non-profit making bodies with legal status such as a society registered under the Societies Registration Act, 1860 (XXI) or under corresponding State Act or trusts registered under the Indian Trust Act or Charitable and Religious Act, 1920, or Co-operative Act or Section 5.25 of the Companies Act, 1956 (Mohanty and Nanda, 1993). But the real problem is that under the 1860 law, anybody can register an organization and engage in a wide range of activities. There is large number of registered organizations in India. It is difficult to identify how many of these are NGOs and how many are not.

The presence of the NGOs is clearly visible in our country today. The international recognition to the valuable contribution of voluntary efforts in India and the resulting international funding to them, have given a boost to the process of voluntarism and as a result, there is a mushrooming of NGOs in India. The role of the NGOs has undergone a change from time to time. A glimpse into the history of the voluntary movements in India is essential for understanding the changing roles of the NGOs in India.

Historical Development

In India, there is a strong link between the history of development organizations and the history of voluntary efforts. The history of social welfare and social reform movements are also linked to the ones mentioned earlier. The religious and the social reform movements like the Brahma Samaj [1], the Arya Samaj [2], the Prarthana Samaj [3], the Ramakrishna Mission [4], the Theosophical movement [5], etc. had a similar approach since they adopted reformist approaches to eradicate the prevalent social evils and the religious dogmas that existed in the society. Untouchability, caste rigidity, human bondage and inhuman treatment of the women and the children were some of the evils, which were deep rooted in the society at that time (Beteille, 1999). Many educational and literary societies were set up, with an aim to expose the emerging middle class to the secular Western thoughts and ideas.

One of the earliest reform movements of the modern type which was greatly influenced by the Western ideas was the Brahma Samaj. Raja Ram Mohan Roy (1774-1883) was the founder of the Brahma Samaj. Roy's contribution to this movement was long lasting. He re-interpreted the Hindu doctrines and found ample spiritual basis for his humanitarianism in the Upanishads. He started a campaign for the abolition of the Sati, a social evil, condemned polygamy and concubinage, denounced casteism and also advocated the right of the Hindu widows to remarry. The Arya Samaj movement could be viewed as an outcome of the reaction to the Western influence. Its founder, Swami Dayanand, rejected the Western ideas and sought to revive the ancient religion of the Aryans (Grover and Sethi, 1970:325-329). The Ramakrishna Mission was founded by Swami Vivekananda in 1897. It was founded with the purpose of preparing the youth to dedicate themselves to the service of God. This could be achieved by serving humanity.

The emergence of the great leader, Mohandas Karamchand Gandhi during the struggle for independence changed the nature of voluntarism in India and there began a process of networking of organizations under his leadership. Gandhiji's call was for collective action to end social evils like untouchability, casteism and illiteracy. People's participation at the grassroots level was essential to achieve this task. Gandhiji felt that 'constructive work' was distinct from 'political work'. According to him, sooner or later, all development work would turn out to be a political work. This was evident from the emergence of such organizations like 'All-India Spinners Association (1925)' and 'All India Village Industries Association (1934)'. [6]

Soon after Independence, Gandhiji made a strong plea for disbanding the Indian National Congress (the political party which came into power), and transforming it into a Lok Sevak Sangh (Public Service Organization), which was rejected (Brass, 1990). Nevertheless, the whole range of 'constructive activities' of the national movement, spearheaded by the Indian National Congress, were continued in the form of 'non-governmental agencies'. It was to these agencies that grants, legal entitlements and charters were made available by the government.

The followers of Gandhi who did not wish to join the government or the ruling party established a number of voluntary (non-governmental) organizations to work closely with the governmental programmes. These programmes were aimed at promoting the welfare of people belonging to diverse social strata, from Harijans (the untouchables) and tribals to slum dwellers, and also for setting up important sectors of a mixed economy, which was the development strategy adopted by the government. Such agencies organized various handicrafts and village industries, rural development programmes, credit co-operatives, educational institutions and retained a degree of autonomy in their functioning (Jain, 1997:125-126).

Later, a new set of social work organizations were added to these, in the rural areas, in community development, panchayati raj and other such bodies. However, there has been a significant change over the years in the character of these organizations. Marcus Franda says, "whether for groups inspired by religion, by Gandhi, by a political ideology or by patronage politics, has run counter to the old ideas of welfare, charity, and social reform to emphasise professionalism, in the service of self-reliant community development" (Franda,

1982:10). The public policy tends to depend heavily on such organizations and as result, it has motivated them to upgrade their technical skills and even to have a regular salaried staff.

The second stage of the growth of the NGOs in India occurred in the 1960s. The iniquitous nature of the State and the inability of the government programmes to benefit the deprived sections, provided the awareness that mere implementation of the governmental schemes by the State-sponsored agencies was not enough, and could in fact be counter productive. This resulted in the emergence of a new set of more struggle-oriented groups. Today, there are a large number of such organizations, struggling for the poor, the landless, the tribals, the bonded labourers, and many others in the social strata that are being discriminated against both by the policies of the State and the dominant elements in the social structure (Chandra, 2001).

These NGOs constitute a whole spectrum of different kinds of voluntary groups. The large NGOs, with their headquarters, their branches and their bureaucracies covering the whole country, are often in close liaison with the ruling groups. They are associated with many developmental schemes, but most often the benefits do not really reach the poor. They seem to suffer from the same weakness that the government bureaucracy and the political parties do. Still they have grown and developed in different sectors. They concentrate on the matters of bonded labour, child labour, women labour, migrant labour, and the labour of those who are forced to work below the subsistence levels. This is the result of a fast changing national and international division of labour (Kothari, 1988:433-453).

This new breed of voluntary organizations has got its own demerits. First of all, they are scattered. Most of them operate only at the micro level; the coordination among them is not strong; and they often find it difficult to work smoothly with each other. Some of them suffer from the lack of leadership, ego complexes and ideological problems. In spite of all these, the positive point is that they work at the grassroots level and seem to be the only link between the poor and the rest of the people (Ibid).

Characteristics and Role of NGOs in India

In recent times, there has been a change in the nature and the character of the voluntary agencies. In the past, such agencies adopted a religious oriented mass approach, in an informal atmosphere. However, at present, the agencies adopt a nationalist oriented group approach, in a formal atmosphere. The objective of the current approach is a socio-economic development of the specified target group, through paid, whole time and formally trained workers. They raise funds from the masses, take interest in government/international aid and collect fees for the services rendered (Padhy, 1986). The earlier approach stressed on the programmes of education, medicine and social reforms in their action plans. The services provided by its members were honorary and free of cost to the beneficiaries.

In the Indian context, it is difficult to make a specific distinction between a voluntary agency and an NGO. An NGO may be a body or an institution registered under the Societies Registration Act, Cooperative Societies Act, Public Trust Act and the Companies Act constituted with a general body, executive council, chief executive, paid staff and volunteers. Such institutions/organizations may concentrate their work at the local, district or state levels. They will specialize in areas like formal education, adult education, and women's education, setting up hospitals and dispensaries and other social service schemes. Around 10,000 NGOs are working in the field of social welfare alone, which got impetus from 1950s onwards, with the establishment of the Central Social Welfare Board. This body provided financial assistance for promoting and strengthening the NGOs in the field of social welfare (Franda, 1982).

Along with that, the introduction of the National Service

Scheme in colleges and the Universities and the establishment of the Union Department of Youth Affairs and Sports played a crucial role in the promotion of voluntary action by the youth and to promote their participation in development. In the 1970s, the voluntary agencies were set up with an aim to implement the developmental projects under the Integrated Rural Development Programme (IRDP). [7] As part of this programme, the educated young families moved to the rural and tribal areas and empowered the rural poor through the NGOs. In spite of this, an analysis of the NGOs working in different fields such as social welfare, rural development and other areas indicate that they are unevenly distributed field-wise. For example, a majority of the NGOs concentrate in the metropolis and very few of these are working in rural and tribal areas (Chowdhary, 1988:4-6).

A number of voluntary organizations have come into existence at the national level during the past two decades, in the field of rural development. These fall into three categories. The first category consists of the techno-managerial voluntary agencies, which work on the principle that the process of rural development can be accelerated through modern management techniques and technology. The second division is the reformist voluntary agencies, which try to bring about changes in the social and economic relationship with the existing political framework; and finally there are the radical voluntary agencies, which seek to challenge the existing production relations. They also undertake some economic, health or educational programmes as an 'entry point' to mobilize masses for political action (Jain, Krishnamurthy and Tripathi, 1985:210). The voluntary organizations and movements started for the purpose of protection of environment like the 'Chipko Andolan' [8] (hug the trees) led by Sunderlal Bahuguna, the Narmada Valley Protection Movement, led by Medha Patkar etc., are also included in this category.

The NGOs have an indispensable role in the development of India. The contribution of the NGOs in India has been substantial in the areas of education—both formal and non-formal, health and medical services, social welfare services, youth development etc. They have been known for their virtues of dedication, flexibility, self-reliance and nearness to the community. However, because of the interaction with the government and the introduction of professionalism in their services, it seems that these virtues have eroded to a large extent. Since the NGOs have started getting the government funds, their flexibility of operations has also diminished. This is because they have to maintain a number of records and registers and engage themselves in much paper work that they have not done before.

In fact, some of the NGOs at the national level are more bureaucratic in their functioning when compared to the governmental agencies. The utilization of the grants calls for rules and regulations which have not only stood in the way of flexibility in their operations, but also initiative, experimentation and self-reliance for which the NGOs were known (Chowdhary, 1985:5). Having such power and patronage, some of the NGOs have become highly politicized and tend to stick to the side of the power elite, than to the deserved people whom they are actually supposed to serve. The politicians have also developed vested interests in these organizations and use them for their political gains rather than helping them to freely accomplish their goals of serving the client groups.

Seventh Five Year Plan (1985-90) envisaged a greater role for NGOs in the implementation of development agenda (Bava, 1997). They are expected to supplement the efforts of the government so as to offer the rural poor choices and alternatives. They are also supposed to become the guides of the people at the village level. They should set an example and become a model. It should be possible for the voluntary agencies to adopt simple, innovative, flexible and inexpensive means with their limited resources to reach a large number of people with less overhead and with greater community participation. The NGOs should activate the delivery system and to make it effective at the village level to respond to the felt needs of the poorest of the poor (Ibid). They should attempt to disseminate information. Besides these, they should strive to make the communities as self-reliant as possible.

The NGOs should also take interest to show how the indigenous resources could be used, how human resources, rural skills and local knowledge which are grossly under-utilized at present could be used for their own development. It is the duty of the NGOs to demystify technology and bring it in a simple form to the rural poor and to train a cadre of grassroots workers who believe in professionalizing voluntarism. They should also take interest to mobilize financial resources from within the community with a view to make the communities stand on their own feet and also to mobilize and organize the poor and generate awareness to demand quality services and impose a community system of accountability on the performance of village level government functionaries (Government of India, 1985:68-70).

The Seventh Plan had given special importance to the role of the NGOs in rural development. An amount of ₹ 150 crore was allocated for the first time, as financial assistance to the voluntary associations in their endeavour to uplift the rural poor. The Government of India had also encouraged the involvement of international voluntary agencies through coordinating organizations like the Central Social Welfare Board and the People's Action for Development (PAD), presently known as the Council for Advancement for People's Action for Rural Technology (CAPART) (Ahmed, 2001).

Human Rights NGOs in India

NGOs have a vital role to play in the promotion and protection of human rights especially in countries like India where the NGO activities are spread in different fields for the welfare of human beings. Resolution Number 38 of the World Conference on Human Rights held at Vienna in Austria, in 1993, recognizes the important role of the NGOs in the promotion of human rights and humanitarian assistance. Human rights' NGOs in India are performing multiple functions related to human rights such as prevention of HIV/AIDS, child exploitation, child labour, bonded labour, training of vulnerable groups, child care, and providing counselling in a number of matters including domestic disputes, the rights of women and children and so on (Kataria, 2013). Human rights NGOs are performing commendable intervention in the following areas which has received national recognition and scholarly appreciation.

One is the social welfare role of the NGOs, where the focus is on relief and charity. Such NGOs can be seen as initiating internal programmes and projects. Another role is the mediatory role, where communication as a skill is important for development and social action. The NGOs which function with this goal can be seen as participating or taking up external programmes and projects. Some NGOs have the consultative role, where documentation and dissemination of information and expertise is critical.

These kinds of NGOs can be seen as working in collaborative programs. Local experts, professionals/resource persons play a major role. Besides these, the NGOs are also involved in development and operation of the infrastructure. The community-based organizations and cooperatives can acquire, subdivide and develop land, construct housing, provide infrastructure and operate and maintain infrastructure such as wells or public toilets and solid waste collection services (Davies, 2014).

Supporting innovation and development of pilot projects is yet another initiative of the NGOs. They have the advantage of selecting particular places for innovative projects and specify in advance the length of time for which they will be supporting the project, overcoming some of the shortcomings that governments face in this respect. They also play a major role in facilitating communication. The significance of this role to the government is that NGOs can communicate to the policymaking levels of government, information about the lives, capabilities, attitudes and cultural characteristics of people at the local level.

The NGOs can facilitate communication upward from people to the government and downward from the government to the people. They also engage in research, monitoring and evaluation. Innovative activities need to be carefully documented and shared—effective participatory monitoring would permit the sharing of results with the people themselves as well as with the project staff (Cousins, 1991). Major national human rights NGOs are People's Union for Civil Liberties (PUCL), People's Union for Democratic Rights (PUDR), Child Rights and You, Bandhua Mukti Morcha, Red Cross, Community Support Concern (CSC), the Centre for Education and Communication etc.

Conclusion

The history of NGOs in India shows that in the initial stage these organizations were not strong enough to protect its desired goals, but gradually they have become strong enough to influence the policies of the Government. NGOs in India have been trying to sensitize, organize and mobilize people at village, local, community level and various other levels, and also among interested groups particularly the poor to enable them to pursue self-development. All these involve a process of education, organization, struggle and reconstruction.

Although NGOs have been quite successful in bringing to the attention of the Government, some of the burning problems in respect of development, poverty, unemployment, illiteracy, ill-health, disparities, incidence of bonded labour, women's rights, peace and social cohesion and communal harmony, equal opportunity to different social, cultural, ethnic and regional groups to pursue development and contribute to national regeneration, yet in the absence of any comprehensive study on the influence of NGOs on the formulation and implementation of policies in India, it is difficult to evaluate their influence on policy processes. Despite this, the impact of various NGOs in certain specific areas is indeed discernible.

For the NGOs to promote local innovations and to come up with solutions to local problems, they must make some of the following features as part of their activities. They should take special interest in building the innovative capacity of the local leaders and the organizations; in providing support to experimentation, documenting new approaches and in spreading these approaches in the areas that are under target.

Besides these, the local NGOs should take steps to provide international acclaim to these innovations. They should make facilities for peer exchange. In addition, striving for the creation of a forum for policymakers to learn about the innovations and incorporating these innovations into their policies should also be one of the missions of these NGOs.

Endnotes

1. Brahmo Samaj is the societal component of Brahmoism, a monotheistic reformist and renaissance movement of the Hindu religion. This movement was started at Calcutta on August 20, 1828 by Raja Ram Mohan Roy and Debendranath Tagore. The first Brahmo Samaj was founded in 1861, at Lahore, by Pandit Nobin Chandra Roy. It literally denotes the community of men who worship 'Brahma' the Creator. In reality, the Brahmo Samaj does not discriminate between caste, creed or religion and is an assembly of all sorts and descriptions of people without any distinction.
2. Arya Samaj is a Hindu reform movement founded on April 7, 1875. It was founded by Swami Dayananda, an ascetic who believed in the infallible authority of the Vedas, and emphasized the ideals of brahmacharya (chastity). He established the "Vedic Schools" or "gurukuls" which put an emphasis on the Vedic values, culture and religion to its students. The students were not allowed to perform traditional idol worship at the school, and were instead expected to perform *sandhya* (a form of meditative prayer using mantras from the Vedas) and participate in *agnihotra* twice daily. All the basic amenities like meals, lodging, clothing and books were given to the students free of cost, and the study of Sanskrit was opened to the non-Brahmins.

The most noteworthy feature of the schools was that only those texts which accepted the authority of the Vedas were to be taught.

3. Prarthana Samaj which meant 'Prayer Society' in Sanskrit, was a movement for religious and social reform based on the earlier reform movements which was started in Bombay. It was founded by Dr. Atmaram Pandurang in 1867. The aim of this society was to make people believe in one God and worship only one God. The main reformers were the intellectuals who advocated the reforms of the social system of the Hindus. The precursor of the Prarthana Samaj in Mumbai was the Paramahamsa Sabha which was a secret society for the furtherance of liberal ideas by Ram Balkrishna Jaykar and others in Mumbai.
4. Ramakrishna Mission is an organisation which forms the core of a worldwide spiritual movement known as the Ramakrishna Movement or the Vedanta Movement. It is a philanthropic, volunteer organisation founded by Ramakrishna's chief disciple Swami Vivekananda, on May 1, 1897. The Mission concentrates on the areas of health care, disaster relief, rural management, tribal welfare, elementary and higher education and culture. It uses the combined efforts of hundreds of ordered monks and thousands of householder disciples. The Mission bases its work on the principles of karma yoga. The Mission has its headquarters at Belur Math in Howrah, West Bengal near Kolkata, in India.
5. The Theosophical Society is an organization which was formed in 1875 in New York City, US., Its main aim was to advance theosophy. It was started by Helena Blavatsky, Henry Steel Olcott, William Quan Judge and others. Its initial objective was the study and elucidation of Occultism, the Cabala etc. After a few years, Olcott and Blavatsky moved to India and established the International Headquarters at Adyar, in Madras (Chennai). They had an interest in studying about the Eastern religions, and these were included in the Society's agenda.
6. An organization called the All-India Spinners' Association was established with the consent of the All India Congress Committee, as an integral part of the Congress organization, but with independent existence and powers. It emerged as a result of the discussion for the establishment of an expert organization for

the development of hand-spinning and khaddar. Such development was not possible without a permanent organization, unaffected and uncontrolled by politics, political changes or political bodies. It had an executive council consisting of the following persons, who held the office for five years—Mahatma Gandhi, Maulana Shaukat Ali, Rajendra Prasad, Satish Chandra Das Gupta, Maganlal K. Gandhi, Seth Jamnalal Bajaj, Shuaib Qureshi,. Shankerlal G. Banker and Pandit Jawaharlal Nehru. The Council had the power to help or open khadi stores to establish a khaddar service, to act as agency on behalf of the Congress to receive self-spun yarn as subscription to the Congress and to issue certificates, and to do all the things that were considered necessary for the furtherance of its objects.

7. The Integrated Rural Development Programme (IRDP) is a rural development programme of the Government of India which was launched in the financial year 1978 and extended throughout India by 1980. It is a self-employment programme intended to raise the income-generation capacity of target groups among the poor. The target group consists largely of small and marginal farmers, agricultural labourers and rural artisans living below the poverty line. Within the target group, there is an assured coverage of 50 percent for scheduled castes and scheduled tribes, 40 percent for women and 3 percent for the physically handicapped. Priority in assistance is also given to the families belonging to the assignees of ceiling surplus land, green card holders covered under the Family Welfare Programme and freed bonded labourers. The objective of IRDP is to provide suitable income-generating assets through a mix of subsidy and credit to below poverty line (BPL) families with a view to bring them above the poverty line.
8. The Chipko Movement was a battle waged by the village folk of Garhwal in Uttarakhand against commercial forestry. Chandi Prasad Bhatt was the founder of the Chipko Movement, which was born in the Alakananda Valley and soon spread to different parts of India. Its main aim was the fostering of love towards trees in the harts of humans. The village folk embraced the trees to prevent others from cutting off the trees and that is how the term 'Chipko', i.e., 'to stick' came into being. In fact, there was a massive participation of the women folk to carry out the protests, which upheld their rights to make use of the forest for

their daily needs of firewood and fodder without damaging the environment. The relevance of the Chipko Movement is that it has led to an awakening of concern for nature, inspiring similar movements in different parts of the country.

References

Ahmed, Naeem (ed.) (2001), "The Challenge of Confidence Building in South Asia: The Role of Civil Society in CBMs in South Asia", New Delhi: HarAnand Publications Pvt. Ltd.

Amnesty International Report (2011), "The State of the World's Human Rights", London: Peter Benensons House.

Amnesty International Report (2012), "The State of the World's Human Rights", London: Peter Benensons House.

Amnesty International Report (2013), "The State of the World's Human Rights", London: Peter Benensons House.

Bava, Noorjahan (1997), "Non-Governmental Organizations in Development: Theory and Practice", in R.B. Jain (ed.), 'NGOs in India: Their Role, influence and Problems', New Delhi: Kanishka Publishers.

Beteille, Andre (1999), "Citizenship, State and Civil Society", *Economic and Political Review*, September 4: 2585-2595.

Brass, Paul (1990), "Politics of India since Independence", UK: Cambridge University Press.

Chandra, Snehlata (2001), "Non-Governmental Organizations: Structure, Relevance and Functions", New Delhi: Kanishka Publishers.

Chowdhary, D. Paul (1988), "Role of NGOs in National Development", *Social Welfare*, November.

Davies, Thomas (2014), "NGOs: A New History of Transnational Civil Society", London: Palgrave Macmillan.

Franda, Marcus (1982), "Voluntary Associations and Local Development in India", Young Asia Publications, New Delhi.

Government of India (1985), "Seventh Five Year Plan, 1985-90", Planning Commission, New Delhi, Vol. 2.

Grover, B.L. and R.R. Sethi (1970), "A New Look on Modern Indian History", S. Chand & Co., New Delhi.

Indian Institute of Human Rights (2007), "Human Rights Activism and Role of NGOs", New Delhi.

Jain, L.C., B. Krishnamurthy and P.M. Tripathi (1985), "Grass without Roots", Sage Publications, New Delhi.

Kataria, S.C. (2013), "Role of NGOs in Protecting Human Rights", New Delhi: Astha Publishers and Distributors.

Kothari, Rajni (1987), "Voluntary Organizations in a Plural Society", The *Indian Journal of Public Administration*, Vol. 33, July-September.

Kothari, Rajni (1988), "NGOs, The State and World Capitalism", *Economic and Political Weekly*, Vol. 21, No. 50, December 13.

Kuriakose, K.K. (1989), "In Quest of Development: Role of Non-Governmental Organisations", Vishwa Yuvak Kendra, New Delhi.

Mohanty, Manoranjan and Biranchi N. Nanda (1993), "NGOs and Funding Centrally Sponsored Schemes", Yuv-Vandhu Cultural Group, Orissa.

Narayana, E.A. (1990), "Voluntary Organizations and Rural Development in India", Uppal Publishing House, New Delhi.

Padhy, Kishore Chandra (1986), "Rural Development in Modern India", B.R. Publishing Corporation, New Delhi.

Shills, D.L. (1968), "Voluntary Associations: Sociological Aspects", *International Encyclopaedia of the Social Science*, Vol. 16, 336-367, Collier-Mac Millan.

The Recommendation of the Commission are Reproduced in AVARD, no.3, Appendix IX.

Index

Index

U

W

Z